VOLUME NINE
Allied Fighters: P-38 series
South & Southwest Pacific 1942–1944

MICHAEL JOHN CLARINGBOULD

Avonmore Books

Pacific Profiles Volume Nine

Allied Fighters: P-38 series South & Southwest Pacific 1942-1944

Michael John Claringbould

ISBN: 978-0-6452469-7-1

First published 2022 by Avonmore Books

Avonmore Books
PO Box 217
Kent Town
South Australia 5071
Australia

Phone: (61 8) 8431 9780
avonmorebooks.com.au

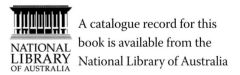 A catalogue record for this book is available from the National Library of Australia

Cover design & layout by Diane Bricknell

Front Cover: A sample of five Pacific Lightnings underlines the multiplicity of markings in the South and Southwest Pacific theatres (top to bottom): F-4 Alice the Goon (8th PRS), P-38G field-modified night fighter (6th NFS), P-38J Tessie (12th FS), P-38F Double Trouble (9th FS) and chequer-tail P-38F #20 (39th FS).

Back Cover: A pair of Ki-43-II Hayabusa conduct a head-on attack against a 9th FS P-38F just south of the Finschhafen coastline in late 1943.

Contents

Michael Claringbould in the passenger seat, alongside Douglas Walker in Michael's M151 jeep in Phnom Penh on 29 January 2013. Walker is the son of General Kenneth Walker, the commander of Fifth Air Force Bomber Command who served under General George Kenney from 3 September 1942 until he went Missing in Action over Rabaul on 5 January 1943.

About the Author

Michael Claringbould – Author & Illustrator

Michael spent his formative years in Papua New Guinea in the 1960s, during which he became fascinated by the many WWII aircraft wrecks which lay around the country and also throughout the Solomon Islands. Michael subsequently served widely overseas as an Australian diplomat throughout Southeast Asia and the Pacific, including Fiji (1995-1998) and Papua New Guinea (2003-2005). Michael has authored and illustrated numerous books on Pacific War aviation. His history of the Tainan Naval Air Group in New Guinea, *Eagles of the Southern Sky*, received worldwide acclaim as the first English-language history of a Japanese fighter unit, and was subsequently translated into Japanese. An executive member of Pacific Air War History Associates, Michael holds a pilot license and PG4 paraglider rating. He continues to develop his skills as a digital aviation artist and illustrator.

Other Books by the Author

Black Sunday (2000)

Eagles of the Southern Sky (2012, with Luca Ruffato)

Nemoto's Travels –The illustrated saga of an IJN floatplane pilot (2021)

Operation I-Go – Yamamoto's Last Offensive – New Guinea & Solomons April 1943 (2020)

Pacific Adversaries Series

Vol One – Japanese Army Air Force vs Allies New Guinea 1942-1944 (2019)

Vol Two – Imperial Japanese Navy vs Allies New Guinea & Solomons 1942-1944 (2020)

Vol Three – Imperial Japanese Navy vs Allies New Guinea & the Solomons 1942-1944 (2020)

Vol Four – Imperial Japanese Navy vs Allies – The Solomons 1943-1944 (2021)

Pacific Profiles Series

Vol One Japanese Army Fighters New Guinea & the Solomons 1942-1944 (2020)

Vol Two Japanese Army Bomber & Other Units, New Guinea & Solomons 1942-44 (2020)

Vol Three Allied Medium Bombers, A-20 Series, South West Pacific 1942-44 (2020)

Vol Four Allied Fighters: Vought F4U Corsair Series Solomons Theatre 1943-1944 (2021)

Vol Five Zero Fighters (land-based) New Guinea & Solomons 1942-1944 (2021)

Vol Six Allied Fighters: Bell Airacobra South & Southwest Pacific 1942-1944 (2022)

Vol Seven Allied Transports: Douglas C-47 South & Southwest Pacific 1942-1945 (2022)

Vol Eight – IJN Floatplanes in the South Pacific 1942-45 (2022)

South Pacific Air War Series (with Peter Ingman)

Volume 1: The Fall of Rabaul December 1941–March 1942 (2017)

Volume 2: The Struggle for Moresby March–April 1942 (2018)

Volume 3: Coral Sea & Aftermath May–June 1942 (2019)

Volume 4: Buna & Milne Bay June-September 1942 (2020)

Volume 5: Crisis in Papua September – December 1942 (2022)

Solomons Air War Series (with Peter Ingman)

Volume 1: Guadalcanal August – September 1942 (2022)

Duel Series (Osprey Publications)

P-39 / P-400 Airacobra versus A6M2/3 Zero-sen New Guinea 1942 (2018)

F4U Corsair versus A6M2/3/4 Zero-sen, Solomons and Rabaul 1943-44 (2022)

P-47D Thunderbolt versus Ki-43 Hayabusa New Guinea 1943/44 (2020)

Introduction

The intention of this volume is to accurately portray a wide cross-section of the Lockheed Lightnings which served in the South West Pacific and South Pacific theatres.[1] Past publications featuring "Pacific Lightnings" have largely ignored Thirteenth Air Force, night fighter and reconnaissance units, many of which are illustrated here for the first time. The mainstay fare for "Pacific Lightnings" in other publications are those of the famous aces, yet the markings regime of this exceptional fighter encompasses a wider and more interesting breadth than generally realised. This volume also takes the opportunity to correct previous incorrect and fictitious profiles.

However, first a caution for Lightning aficionados. Much existing material about the role and history of the type in the Pacific is questionable or erroneous. The Lightning's actual operational history reflects an alternative version to the many well-publicised accounts of its exceptional combat prowess. The statistics are unflattering: a total of 939 Lightnings served in both Pacific theatres from April 1942 to May 1944, ranging from the F-4 reconnaissance version through to the natural metal finish P-38J-20 model (about 300 additional Lightnings were later added to the Philippines theatre inventory, which are outside the scope of this volume). Of these 939 airframes, 192 were lost to combat, 37 to weather-related causes, and 171 to various other causes, leaving 539 to either proceed to the Netherlands East Indies and Philippines theatres or be left behind. In such cases they either served with the Combat Replacement Training Center or were scrapped at Nadzab or Finschhafen. The fact is that a Lightning pilot in the Pacific had a roughly equal chance of losing his life to tropical weather or an accident than being shot down by the enemy. Furthermore, many non-combat losses resulted directly from asymmetric engine-out situations, mostly on take-off or landing, a problem less lethal in comparable single-engine fighters.

The amount of Fifth Air Force Lightning losses during the Rabaul campaign in October/November 1943 is instructive. The biggest Rabaul mission by the Fifth was "Bloody Tuesday" of 2 November 1943, during which Mitchells claimed 26 Japanese fighters and the escorting Lightnings a further 29, for a total of 55 American kills. Yet the true American score was only fourteen Zeros, thus they claimed nearly four times as many as actually shot down. In return the Americans paid a terrible price losing eleven Mitchells and nine Lightnings, a combined total of twenty aircraft. Thirteenth Air Force Lightning losses during the January/February 1944 Rabaul campaign are dreadful. Indeed, a sobering examination of specific Zero versus Lightning combats reveals the Zero still retained a slight edge, a statement likely to be met with disbelief in many circles. The popular and accepted version of Lightnings giving quick demise to multifarious Zeros over Rabaul in this timeframe is illusory.

---

1   The South West Pacific Area as defined by this volume essentially means the New Guinea theatre, between Dutch New Guinea in the west and the South Pacific theatre in the east. For this reason, the three F-4 reconnaissance Lightnings operated by the RAAF are outside the scope of this volume as they were used over the Netherlands East Indies. A few other Lightnings were loaned to the RAAF in New Guinea but were operated by the 8th PRS, USAAF.

Neither is it true that the Japanese thought the Lightning particularly dangerous, and neither did they give it any nickname as often claimed. In fact, the diaries of several JAAF Wewak Ki-43 pilots record a preference for tackling the type at lower altitudes, where its clumsiness in a low energy state made it easy prey for a pilot with good manoeuvrability. To IJN pilots the Lightning was yet another adversary they preferred to engage at lower altitudes and on their terms. This, of course, was not always possible. Whilst true that the Lightning possessed higher speed compared to its Japanese opponents, some historians questionably claim this bestowed the type as overall "markedly superior" to its adversaries.

It is also true that the range of the P-38 was better than the single-engine fighters available in the Pacific throughout 1942/43, making it possible to engage the enemy over distant locations such as Rabaul. Largely for this reason, Fifth and Thirteenth Air Force commanders kept pressing Washington for more Lightnings. The commanders were supported by their pilots who considered it the fighter of choice.

However, we need to examine how and why the Lightning performed so poorly in the Pacific. Every new type has its teething problems, but the ones faced by the Lightning were second only to the Corsair, technologically even more advanced. The Lightning, a huge airframe, was delivered to a backward theatre but the airframe itself portended advanced technology. This contradiction quickly established time-consuming setbacks, substantially delaying its introduction to combat. The most glaring hindrance was leaking fuel tanks, to which field repairs proved especially time-consuming, and delayed the type's introduction to combat. The problem was rectified in later models, but the leaks were persistent in both the P-38F and P-38G models. Compared to single-engine fighters, the enormous airframe demanded about four times the man-hours to keep them airworthy. Aside from an extra engine to maintain, advanced ancillary systems such as control servo boosters required nuanced technical skills. Supercharger coolers ran hot, requiring pipes to be extended in the field to avoid splitting. Machine-gun recoil frayed gun-mount bearings and the guns eventually required steel tie-downs. Heavy rain played havoc with the electrical systems including shorting out solenoids and electrical plugs. Wing-mounted drop-tanks damaged the flaps when released in combat, a problem solved by adding fins to the tanks.

The most questionable claim is that its twin engines offered an additional safety factor, so essential for the lengthy stretches of ocean and jungle unique to the Pacific. However, once again the contention is not substantiated in operational records, and in fact they suggest the reverse. The type's contra-rotating V-1710 Allison in-line engines were glycol-cooled, thus more susceptible to combat damage than robust radials such as those which powered the P-47D. Two engines meant almost twice the chance of having an engine hit in combat. Losing an engine was usually lethal on take-off. Asymmetric thrust was a challenge to handle even in cruise, but particularly so in instrument conditions or tough weather. A single-engine Lightning was a sitting duck for Japanese fighters. There are numerous times when Lightnings were written off back at base when trying to land with one engine operative. An asymmetric low-level approach with flaps down on one engine was always an unwanted proposition. Many pilots did not live to talk about it.

An alarming sign of things to come was when the 17th FS Squadron (Provisional) lost seven P-38Fs during its first few weeks of training and test flights in Australia, the first loss occurring at Amberley on 7 September 1942. Perhaps symbolic is the fate of the most famous Fifth Air Force Lightning, *Marge*, which in 1944 was allocated to the USAAF's highest-scoring pilot, Richard Bong. It was lost, not in combat but due to engine and miscellaneous systems failures (see Chapter 22).

The Lightning's manoeuvrability was markedly inferior to that of more nimble Japanese opponents, and enemy fighter units fought accordingly where possible. Avoiding combat at low altitudes and the use of fast diving attacks were obvious P-38 tactics, however, claims that these enabled the P-38 squadrons in both New Guinea and the Solomons to achieve superior results are incorrect. As a result of these tactics Lightnings often found themselves at low altitude in what is termed aeronautically as a "low energy state". The Zero and even the lightly armed Ki-43 found the Lightning easy prey in such a condition. It is true that when the Lightning's unique performance characteristics were maximised, the results could be devastating, but then the same could also be said of Japanese fighters. The wise Lightning pilot could always use his outstanding speed to escape combat, but an obvious corollary is that such action did not shoot down enemy aircraft.

Perhaps the biggest myth which continues to be perpetrated is the alleged favourable kill ratio of the Lightning against the Zero. Armed with accurate and detailed Japanese unit logs, right down to the number of rounds fired, actual combat results instead of readily accepted USAAF claims give another reality. Both US and Japanese pilots over-claimed by about three to one, or even more in many cases. This means, on average, reducing all Lightning combat claims by about two thirds. As one example, 38 pilots from the 475th FG (which exclusively operated P-38s), enjoy "ace" status however the true number is closer to thirteen. If you really want to nail down an individual pilot's score it is easy to do so – the Japanese records offer a ready comparison of claims versus reality. In particular, the early claims by the 39th FS in its first few months of combat were almost frenetic.

In the South Pacific theatre combat engagements were often fought by multiple units and often different services were involved, with no cross-checking on combat claims. A good example is "Ace-in-a-Day" 339th FS pilot Lieutenant Murray Shubin. Awarded five Zeros on 16 June 1943 over Guadalcanal, Shubin's own claims were two definites and four probables, and by his own admission he witnessed none of the probables crash. Instead, he was awarded the extra kills upon evidence from an army ground observer miles away from the action, an observation readily accepted by Thirteenth Air Force Command, perhaps driven more by prestige than realism. Such a witness would have found it challenging to discern Shubin's Lightning from the many others airborne, let alone confirm which particular victim was his kill. Total awards between USAAF, USMAC and USN aerial units for this particular engagement were 49 Zeros, yet only fourteen actually fell. How then to accurately and fairly reapportion these kills?

None of the above detracts from the uniqueness of the Lightning or the courage of those who flew it. Its matchless range enabled it to take the fight to the enemy, the Hollandia campaign in particular being a beneficiary. The role of the type in the Pacific is incomparable to other theatres due to the harshness of the climate and the fearsome geography of New Guinea and the Solomons. Many Lightnings were swallowed in this tropical maw, and many are still missing today. Its unique design warranted unique markings, which I hope you enjoy, happily obliged to us courtesy of the Fifth and Thirteenth Air Forces.

Michael John Claringbould
Canberra, Australia
July 2022

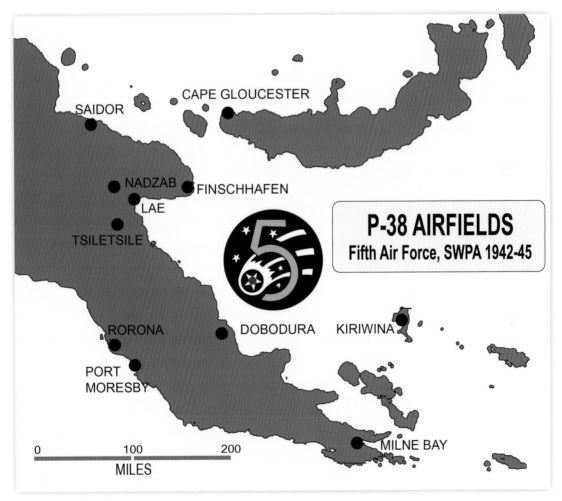

A map showing key airbases used by Fifth Air Force P-38s in the New Guinea area of the South West Pacific Area command during 1942-1944. Note that some of these locations had several airfields. For example Port Moresby hosted six sizeable airfields: Three-Mile (Kila), Five-Mile (Wards), Seven-Mile (Jacksons), 12-Mile (Berry), 14-Mile (Schwimmer) and 17-Mile (Durand).

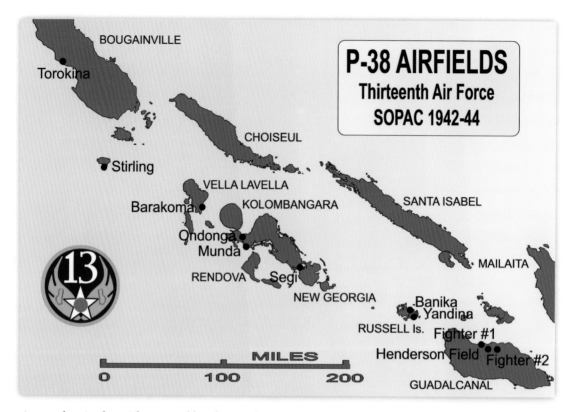

A map showing key airbases used by Thirteenth Air Force P-38s in the Solomons area of the South Pacific Area command during 1942-1944.

Glossary and Abbreviations

CN	Constructor's Number
CRTC	Combat Replacement Training Center
FG	Fighter Group
FS	Fighter Squadron
IJN	Imperial Japanese Navy
JAAF	Japanese Army Air Force
NFS	Night Fighter Squadron
POW	Prisoner of War
PS	Pursuit Squadron
PRG	Photographic Reconnaissance Group
PRS	Photographic Reconnaissance Squadron
RAAF	Royal Australian Air Force
RCAF	Royal Canadian Air Force
RNZAF	Royal New Zealand Air Force
SOPAC	South Pacific Area
SS	Steamship
SWPA	South West Pacific Area
US	United States
USS	United States Ship
USN	United States Navy
USAAF	United States Army Air Force

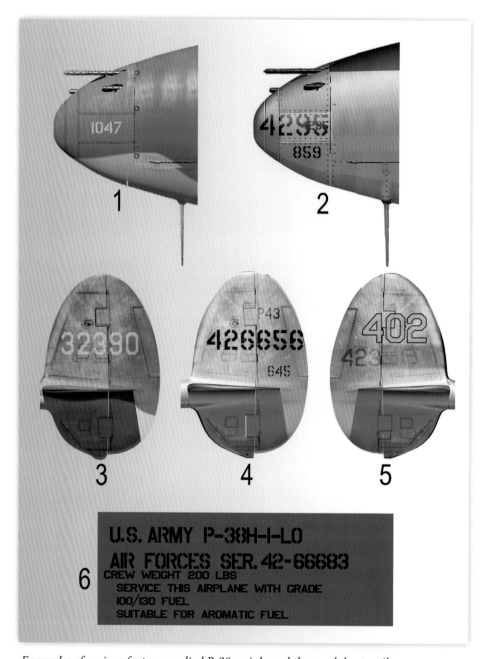

Examples of various factory applied P-38 serials and the gondola stencil.

CHAPTER 1
P-38 Markings

Several factory markings on the P-38 series are unique and can assist in identifying the aircraft's serial number. All aircraft carried what was termed a four-digit Constructor's or Manufacturer's Number (CN or MN) on the nose. This was stencilled in either yellow or black depending on whether it was applied over Olive Drab or natural metal finish. Lockheed also issued batch numbers with serial number production runs: thus P-38G serial 43-2185 was CN 3294, but in Lockheed terms it was CN 322-3294, although this three-digit prefix was not stencilled on the airframe. Tail serial numbers were also stencilled in either yellow or black depending on application over Olive Drab or natural metal finish.

On later batch runs of P-38Js, on the nose either the last three digits of the USAAF serial number were stencilled over the CN, or in other batch runs instead the CN was replicated on top of the smaller one in larger stencils.

Some examples of P-38 markings are shown on page 12:

- Example 1 shows the location of where the CN was stencilled on all Olive Drab models, with slight variations in position depending on the workman of the day. Some Olive Drab batches stencilled the last two numerals of the CN in larger stencils over the four-digit stencil.

- Example 2 depicts factory stencils for P-38J-20 serial 44-23291. The digits 4295 are the CN, applied over a smaller 4295, and not the last four digits of the serial number. The smaller stencil 859 is a factory order number.

- Example 3 shows the tail serial number stencilled in yellow over an Olive Drab airframe. All P-38F, P-38G and P-38H models carried the stencil size and style as shown. Note that the numeral "2" is slightly altered from the official USAAF specification.

- Example 4 shows the tail serial number style used for the P-38J-15 series onwards. Lockheed sub-contracted out several sub-assemblies including rudders. Hence P43 is a sub-contractor part number and 645 defines the 645th sub-assembly made of this particular part.

- Example 5 shows the effect of natural metal finish Lightnings having their serial numbers removed in the field by rubbing over with emery cloth. This explains why on some profiles, the serial or parts thereof, remain partially visible as shown here. Factory serials were similarly erased on the nose of natural metal finish fighters. Residual black paint retained in the cloth often stained areas of the removed serial.

- Example 6 shows the layout of the manufacturer's serial block stencil which appeared on the left-land side of the gondola. This style and wording remained constant throughout all production models.

The subject of Profile 1, P-38F serial 42-12742, at Amberley shortly after its assembly in early September 1942.

P-38Fs after assembly at Amberley near the 17th FS (Provisional) camp area.

CHAPTER 2

First Lightnings: 8th PRS and 17th FS (Provisional)

The first four examples of Lockheed's radical new design of the twin-boom Lightning arrived in Australia on 9 April 1942 as deck cargo on the former luxury liner SS *President Coolidge*. They were F-4 reconnaissance versions of the P-38, painted in a haze-blue scheme, and were assembled and test flown at the Commonwealth Aircraft Factory located at Port Melbourne. All four were assigned to A Flight of the newly arrived 8th Photographic Reconnaissance Squadron, and by the start of May two F-4s had been rushed north to Townsville.

This was on the eve of the Battle of the Coral Sea and both aircraft were urgently needed for reconnaissance missions over New Guinea. However, one vanished on its first mission and the other was badly damaged during a crash-landing at Townsville. The 8th PRS did not resume regular missions until July 1942 (see Chapter 5).

Meanwhile orders for the 39th FS to transition from Airacobras to the twin-engine Lightning fighters had been issued on 1 July 1942 but it was not until 18 October 1942 that the first P-38s were stationed in the front line, at Port Moresby's 14-Mile airfield (see Chapter 3).

The first batch of 49 P-38s had arrived in Brisbane as deck cargo on the liberty ship SS *John Wise* on 12 August 1942. These P-38Fs were assembled at RAAF Amberley, before being test flown by a unit designated as the 17th Fighter Squadron (Provisional), also known as the 17th Provisional Fighter Squadron.[1] Prior to delivering the P-38s to the 39th FS at Townsville, a few weeks of test flying and training saw the 17th FS (Provisional) lose seven P-38Fs to operational causes. These losses reflected the increased challenges in operating the demanding new type, with the first loss occurring at Amberley on 7 September 1942.

P-38F number 16 taxies at Amberley shortly after assembly in early September 1942.

---

1   The 17th Pursuit Squadron (Provisional) had originally been formed in Australia in January 1942 and in subsequent weeks flew P-40Es in combat in Java. However, it was never formally raised by the USAAF and only existed for a short time. The original 17th PS had been a pre-war P-40E unit in the Philippines, elements of which did not surrender until May 1942. The re-use of this designation (again for a relatively short time) for the introduction of P-38s in the SWPA reflects a strong historical link to the pre-war Far East Air Force in the Philippines, the fighters of which were organised under V Interceptor Command. This lineage was reflected in the Fifth Fighter Command, Fifth Air Force, which was established in the SWPA in September 1942 (see Chapter 22).

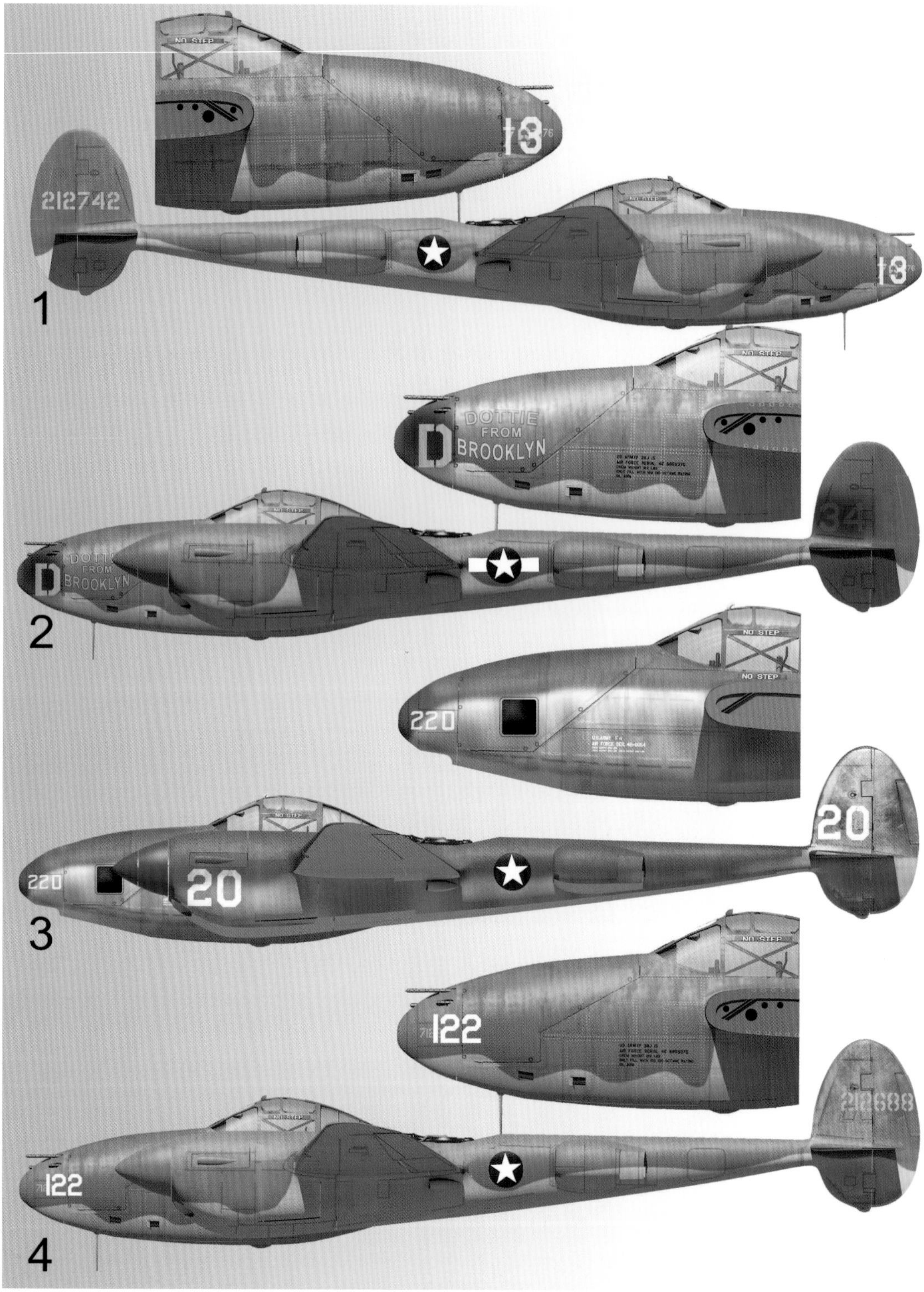
1
212742
NO STEP
2
DOTTIE FROM BROOKLYN
D
34
3
220
20
4
122
212688

Profile 1– P-38F serial 42-12742, 17ᵗʰ FS (Provisional), squadron #13, Port Moresby, 21 September 1942

This airframe was among the first batch of P-38Fs which arrived at Brisbane on 12 August 1942. It was assembled at RAAF Amberley and subsequently test flown by the 17ᵗʰ FS (Provisional). It was the thirteenth airframe assembled and thus given squadron number 13. It is portrayed as it appeared during its navigation exercise to Port Moresby via Horn Island on a familiarisation flight of 21 September 1942. This fighter later served with the 39ᵗʰ FS and was decommissioned in late 1943. Note the Constructor Number 7176 which appears in yellow stencils under the squadron number 13 on the nose. Behind it are the last two digits of the Constructor's Number in larger yellow stencils.

Profile 2 – P-38F serial 42-12647, 27ᵗʰ Air Depot, *Dottie From Brooklyn*, Port Moresby, January 1944

This Lightning was also among the first batch to arrive in Australia and was later assigned to the 39ᵗʰ FS as squadron number 34. Named *Dottie From Brooklyn* by Lieutenant Wayne Rothgeb after his wife, on 14 May 1943 Rothgreb had to manage a shattered starboard turbocharger at 27,000 feet, forcing him to return to Schwimmer 'drome, Port Moresby. Following repairs, the fighter was reassigned to 80ᵗʰ FS pilot Lieutenant Cornelius "Corky" Smith and assigned the squadron identifier "D", and the original name was retained. On the morning of 21 June 1943 Smith was credited with three kills following combat with 24ᵗʰ *Sentai* Ki-43-IIs near Wau, although only one Japanese fighter was lost. Shortly afterwards the Lightning was considered too worn for combat and was transferred to the 27ᵗʰ Air Depot at Port Moresby where it was used as a hack. During a flight around January 1944 both engines lost power and it force-landed into marshland near the village of Lea Lea. The fighter was recovered in 1978 and sent to Australia. The fighter is illustrated as it appeared on the day it was lost.

Profile 3 – F-4 serial 41-2220, 8ᵗʰ PRS, squadron #20, 14-Mile 'drome, December 1942

This airframe was received by the Sacramento Air Depot in October 1942 from an unidentified training unit. It was subsequently shipped to Townsville where it was reassembled and then flown to Port Moresby on 29 November 42 for use by the 8ᵗʰ PRS. In August 1943 it was lent to No. 75 Squadron, RAAF, for five months along with another F-4 (serial 41-2156). On 21 December 1943 both F-4s were returned to the 8ᵗʰ PRS, with which they continued service in New Guinea, until decommissioning in September 1944.

Profile 4 – P-38G serial 42-12688, 17ᵗʰ FS (Provisional), squadron #122, Townsville, September 1942

This Lightning was the first "G" model assembled and assigned into the Fifth Air Force. It was given the squadron number 122 which represented the last three digits of the Constructor's Number 7122. A limited number of similar squadron numbers were carried into combat with the 39ᵗʰ FS as late as December 1942 after which they were replaced with two-digit 39ᵗʰ FS numbers in the 10 to 39 range.

The subject of Profile 8, P-38F serial 42-12621, at Kiriwina in late 1943 with an RAAF refuelling truck.

Captain Charles Gallop with the subject of Profile 9, P-38F serial 42-12627 Loi, at 14-Mile (Schwimmer 'drome) in March 1943.

CHAPTER 3
39th Fighter Squadron

It was the 39th Fighter Squadron which first took the Lightning to combat in the Pacific. Major George Prentice was appointed the new squadron commander on 18 September 1942 with the express task of transitioning the unit from the Airacobra to the P-38. The unit receives disproportionate coverage in this volume because of its significant status as the first user of the new type.

Prentice led twelve Lightnings from Amberley to Townsville on 21 September 1942, where the airframes and guns were checked for combat-ready status. The next morning, they flew on to Seven-Mile 'drome in Port Moresby where they were briefed that the 39th FS would be based at newly completed 14-Mile airfield. The dozen P-38Fs then made a brief five-minute ferry flight to their new base where they settled into a tented camp already established by the squadron's advance element. By month's end the squadron had reached its full P-38 allocation. Over the next few weeks, the remainder of the squadron's pilots moved up to New Guinea from Australia, maintenance and training schedules permitting.

However, introduction of the new type to combat proved problematic. Engineers soon found that the new fighter's advanced technology presented complex and demanding setbacks. Compared to the squadron's previous mount, the Airacobra, the Lightnings needed about four times the man-hours to keep them airworthy. Not only was there an extra engine to maintain, the complicated airframe also housed advanced systems such as control servo boosters. This required additional skills for maintenance personnel to master before other problems soon started appearing. The pipes to the supercharger coolers were too short and had to be extended to avoid splitting. Machine gun recoil brindled the gun mount bearings, requiring steel tie-downs. Furthermore, when long-range drop-tanks were released at high speed they damaged the flaps, a problem solved by adding guide fins. The most challenging problem, however, proved to be leaking fuel tanks, to which repairs proved finicky and time-consuming. This hindrance was rectified in later models, but the leaks created no end of frustration during the first months at Port Moresby.

From 14-Mile the 39th FS initially flew a handful of uneventful escort missions. The Lightnings would not see combat until late November 1942 as in the interim the squadron sent a temporary detachment to Guadalcanal. The request came on 29 October when eight Lightnings were ordered to Milne Bay from where they would conduct a 600-mile ferry flight to Guadalcanal guided by a B-17E. The bomber was also tasked to ferry engineers and spare pilots, but the bomber arrived two days too late to find the operation cancelled. A second request saw eight P-38Fs fly to Henderson Field on Friday 13 November via Milne Bay, only one day after the arrival of the first SOPAC Lightnings (see Chapter 15). The following day the Lightnings strafed beached Japanese merchant ships on Guadalcanal's north shore, then flew uneventful patrols and escort missions. Seven Lightnings returned to 14-Mile on 22 November as the eighth was grounded pending parts and repairs.

On 14 November 1942 Lieutenant Richard Ira Bong, the future Fifth Air Force leading ace, was temporarily assigned to the 39th FS to get combat experience with the Lightning. Subsequently the 39th FS operated P-38s for just over a year before commencing transition to P-47Ds in November 1943, with their P-38 inventory being transferred to the 475th FG. During this period the squadron lost 33 Lightnings to non-combat causes, with an additional three missing and another dozen lost to combat, the last on 4 December 1943.

Markings

The 39th FS at first operated mainly P-38Fs in the serial 42-12625 to 42-12859 range, and later acquired G and H model replacements. The squadron's signature marking was the squadron number on the nose and tail painted or stencilled in white using modified USAAF serial number calligraphy. Shark teeth soon appeared on some cowls, sprayed over a thin aluminium sheet template. Later versions of the shark's teeth had red applied to the interior of the teeth and were sometimes highlighted with a black outline. Several aircraft were named, however there was no dedicated squadron artist at this early stage, so nose-art was modest and minimal. Several pilots personally decorated their own aircraft. Squadron numbers were applied in the 10 – 39 range and serial numbers on the tail were usually painted over before applying the white squadron number.

Lieutenant Curran "Jack" Jones poses with the subject of Profile 8 at 14-Mile 'drome, Port Moresby, in early 1943.

The 39*th* FS commander Major Thomas Lynch (circled) poses with the subject of Profile 11, P-38G serial 42-12715, and other pilots at 14-Mile in May 1943.

Lieutenant Lee Haigler about to climb aboard the subject of Profile 5, P-38F serial 42-12666, at 14-Mile 'drome. During a squadron scramble he has just jumped off a racing vehicle carrying several other pilots.

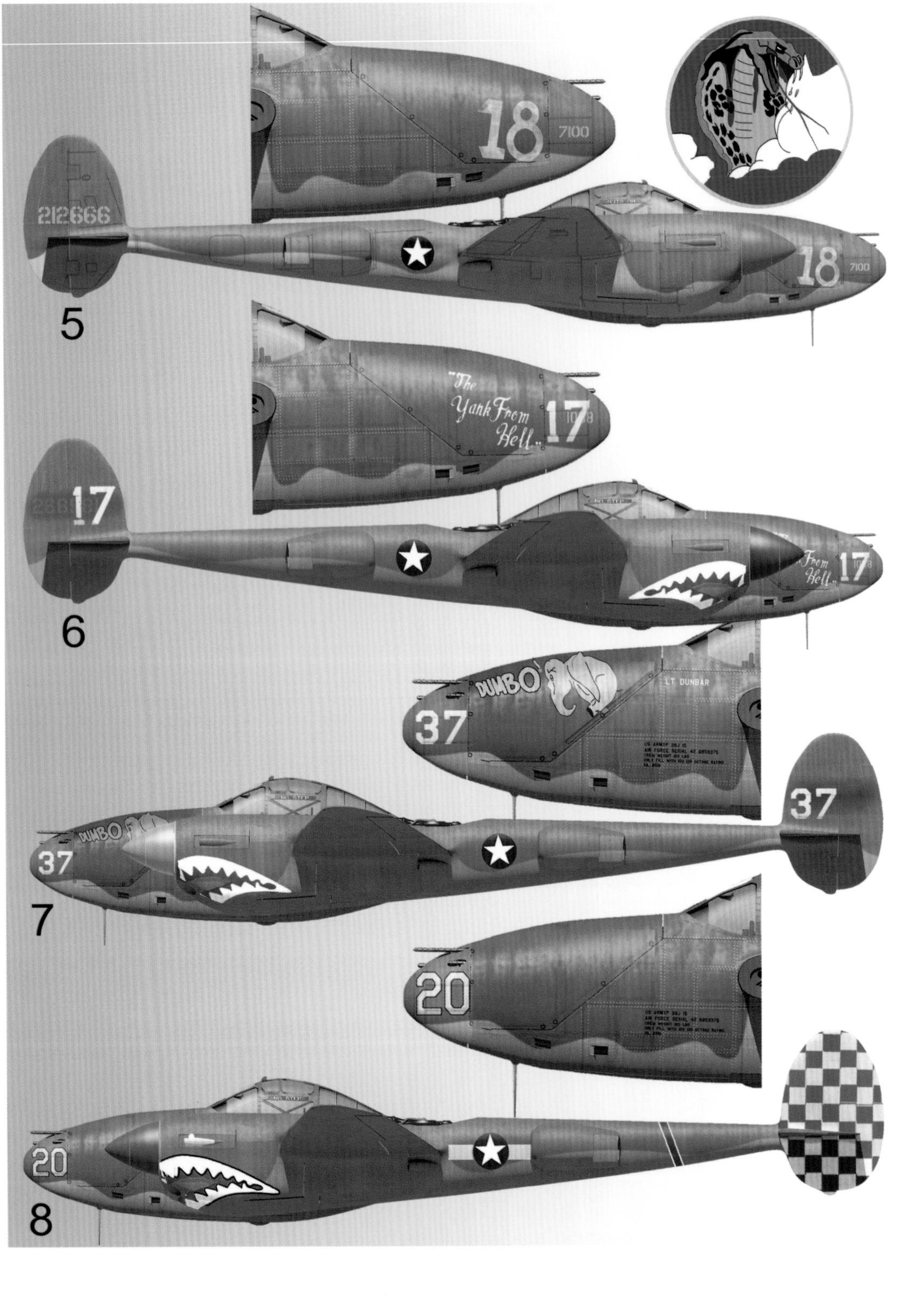

212666
18
7100
5
"The Yank From Hell"
17
1078
6
DUMBO
LT DUNBAR
37
7
20
8

39ᵗʰ FS Logo –The logo was a cobra in the clouds, deriving back to the unit's P-39 Airacobra days.

Profile 5 – P-38F serial 42-12666, squadron #18, 14-Mile 'drome, October 1942

This Lightning was among the first batch of 33 P-38Fs delivered to the 17ᵗʰ FS (Provisional) at Amberley. It was reassigned to the 39ᵗʰ FS at Townsville in October and arrived at Port Moresby later that month. For the first few weeks this early batch had the squadron number painted on the nose, leaving the serial number intact on the tail. The factory applied CN 7100 appeared in yellow stencil on the nose. Note that the hand-painted number 18 is slightly offset, applied when the stationary airframe on the ground had an upwards angle. This fighter was written off in an accident on 2 February 1943.

Profile 6 – P-38H serial 42-66581, squadron #17, *The Yank from Hell*, 14-Mile 'drome, November 1942

This squadron number 17 replaced the first "17" which crashed at Amberley on 4 December 1942 during flight training, a P-38F named *Synchronized Sal*. Assigned into the 39ᵗʰ FS around mid-1943, this P-38H was mostly flown by Lieutenant Edward Flood who named it. Part of the yellow stencil CN 1088 remained visible on the nose when over-painted with the squadron number, and the serial number on the tail remained partially visible too. The fighter was later transferred to the 431ˢᵗ FS with which it was written off during a wheels-up landing at Kiriwina following the infamous 2 November 1943 Rabaul raid.

Profile 7 – P-38F serial 42-12847, squadron #37, *Dumbo*, 14-Mile 'drome, July 1943

This fighter was assigned to Lieutenant John Dunbar, who arrived at Port Moresby in September 1942 as one of 26 new pilots who had already undertaken P-38 training in the US. Dunbar decorated the aircraft himself on 9 July 1943 shortly after it was reassigned to him as a first lieutenant. As a result of the nose art Dunbar himself soon acquired the nickname "Dumbo" among his squadron mates.

On 2 September 1943 the fighter force-landed at the emergency coastal airfield at Terapo flown by Lieutenant Hamilton Laing who put down with an engine out and low fuel. Dunbar was assigned an unnamed replacement fighter: P-38H 42-66911, squadron number 37, which was later lost with another pilot over Rabaul on 7 November 1943. As to the original *Dumbo*, after six decades it was salvaged from Terapo in 2002 and shipped to Australia where it was undergoing restoration at the time of publication.

Profile 8 – P-38F serial 42-12621, squadron #20, Seven-Mile 'drome, July 1943

This fighter was often flown by Lieutenant Stanley Andrews when it was first assigned into the 39ᵗʰ FS at the end of 1942. It was destroyed on 1 August 1943 at Port Moresby during a landing accident when being flown by Lieutenant Curran "Jack" Jones. The fighter is portrayed as it appeared just before it was lost, with the white bars added to the star in July 1943, a unique Fifth Air Force marking. The fin and rudder assemblies were painted in blue and white stripes sometime around mid-1943 which denoted Jones as a flight leader.

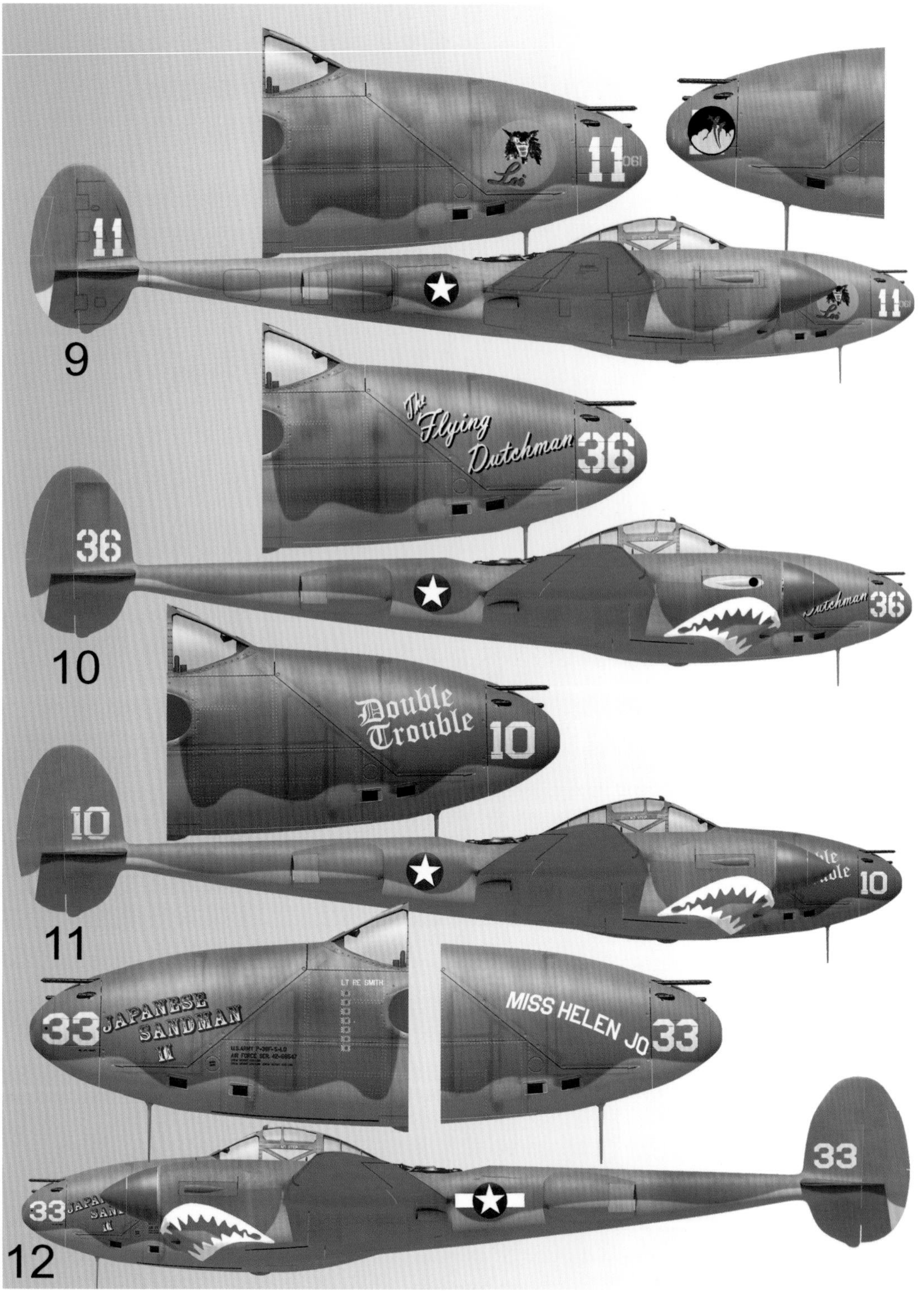
11
9
The Flying Dutchman
36
Dutchman
10
Double Trouble
10
11
33
JAPANESE SANDMAN II
LT RE SMITH
MISS HELEN JO
33
12

Profile 9 – P-38F serial 42-12627, squadron #11, *Loi*, 14-Mile 'drome, March 1943

Mostly flown by Captain Charles Gallop, this fighter was destroyed in an operational accident on 24 July 1943. Gallop named the fighter however the meaning of the name *Loi* is unclear. A circular motif of a man riding two jumping sharks was applied to the port nose and, unusually, applied over the squadron number.

Profile 10 – P-38F serial 42-12645, squadron #36, *The Flying Dutchman*, 14-Mile 'drome, March 1943

This fighter was commonly flown and named by Lieutenant EW Randall. It was lost to an accident in April 1943.

Profile 11 – P-38G serial 42-12715, squadron #10, *Double Trouble*, 14-Mile 'drome, March 1943

This airframe was the first assigned to 39th FS commander Major George Prentice who chose the squadron number 10 and also named the fighter. This P-38 was later shot down by Japanese fighters on 3 March 1943 during the Battle of the Bismarck Sea when flown by Lieutenant Fred Shifflet. After Prentice became the commanding officer of the 475th FG in May 1943 he named all his subsequent P-38s similarly.

Profile 12 – P-38H serial 42-66905, squadron #33, *Japanese Sandman II*, Dobodura, September 1943

This Lightning was assigned to Lieutenant Richard Smith in mid-April 1943 who named it *Japanese Sandman II* on the port nose and *Miss Helen Jo* on the starboard. In late 1943 when the 39th FS transitioned to the P-47D, the fighter was transferred to the 431st FS and continued to operate from Dobodura. On 4 December 1943 it departed Dobodura flown by Lieutenant Dolphus Dawson for a training exercise. Shortly after take-off Dawson experienced mechanical failure and put down into a swamp not far from Dobodura. The fighter is illustrated as it appeared at the end of September 1943 after Smith had applied his seventh victory flag, when he claimed a Ki-43 fighter on 23 September 1943.

The stripe on the boom of number 27 denotes a flight leader, in this case Lieutenant Charles King. The fighter is P-38F serial 42-12653, seen at 14-Mile around March 1943.

Profile 13 – P-38H serial 42-66532, squadron #10, Dobodura, September 1943

This Lightning was assigned to 39th FS commander Major Thomas Lynch who replaced George Prentice on 24 March 1943. Its blue tailplane tip and double blue piped stripes indicated his position as commander; however, the fighter was lost on 13 November 1943 with Lieutenant Lee Haigler. While escorting F-5 reconnaissance Lightnings to Rabaul at 30,000 feet, Haigler was last seen in bad weather around 40 miles north of Buna.

Profile 14 – P-38F serial 42-12623, squadron #16, 14-Mile, March 1943

Assigned to Lieutenant Robert Faurot, this fighter went missing on 3 March 1943 during the Battle of the Bismarck Sea, having been shot down by Japanese fighters. The motif on the unnamed fighter portrays a skull with glasses wearing top hat, with an ace of spades on the collar.

The subject of Profile 12, P-38H serial 42-66905 Japanese Sandman II, at Dobodura.

The subject of Profile 14, P-38F serial 42-12623, taxies over Marston matting at 14-Mile in early 1943.

The subject of Profile 15, P-38H serial 42-66903, in flight over the Solomons.

CHAPTER 4
6th Night Fighter Squadron (Dets. A & B)

The 6th NFS operated night fighters in both the SOPAC and SWPA theatres split into Detachments A and B. The first served with the Fifth Air Force in New Guinea from February 1943 for eight months equipped mainly with the Douglas P-70, but was recalled to Hawaii at the end of the year. During its time at Port Moresby it converted two ex-80th FS P-38Gs to night fighters by adding SCR-540 radar units and installing an extra machine gun and flash suppressors. The New Guinea detachment relied heavily on the 80th FS for logistics and maintenance facilities at Port Moresby.

Detachment B, also mainly operating P-70s, operated from Guadalcanal where it arrived on 28 February 1943 and received a handful of used P-38Hs from the 339th FS, along with several 339th FS pilots including Lieutenant Henry Meigs. These were equipped with radar to perform as single-seat night fighters operating from Henderson Field, designed to curb the activities of Japanese nuisance sorties over Guadalcanal. The airframe conversions mainly accommodated different radios and instruments.

Operating at Guadalcanal in conjunction with a RNZAF radar unit (callsign "Kiwi"), the Guadalcanal Detachment soon discovered that when a P-70 approached a Japanese intruder, the enemy could outrun the attacker by diving to accelerate. An effort was made to equip the Lightnings with the USN AN/APS–4 airborne radar, but the initiative was diminished because of the excessive workload imposed by single-pilot instrument flying at night. As a result, tactics were changed so that P-70s patrolled an outer zone with radar coverage while the 6th NFS's P-38Hs circled in a nearby inner zone aided by searchlights. Loitering overhead Guadalcanal at altitudes up to 30,000 feet, the P-38Hs relied on searchlights to illuminate the enemy, a reliance which gave Meigs two kills in the early hours of 21 September 1943. The operational logs of No. 702 *Ku* confirm that Meigs downed two of six Bettys conducting a nuisance raid from Buka.

When the 419th NFS later took up station on Guadalcanal on 15 November 1943, and was assigned into the 18th FG, it assimilated the 6th NFS inventory in the process. Detachment B was disbanded on 15 December.

Markings

Detachment B in the Solomons repainted the bottom half of the P-38H airframe matt black and applied the last three digits of the Constructor's Number as the squadron identifier, directed in yellow stencil on the forward gondola and the air scoops on the booms. The top half of the fins were also painted matt black. Detachment A in New Guinea painted their P-38Gs overall matt black.

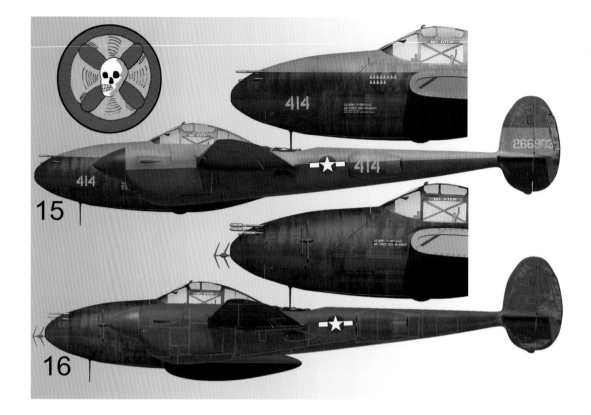

6th NFS Logo

The logo was a skull in front of a revolving propeller.

Profile 15 – P-38H serial 42-66903, squadron #414, Guadalcanal, September 1943

This fighter was transferred from the 347th FG, although from which squadron is unknown. Note each night mission was given a yellow marker. The squadron number 414 is the last three digits of the fighter's Constructor's Number 1414.

Profile 16 – P-38G serial unknown, 12-Mile 'drome, Port Moresby, July 1943

This overall matt black Lightning was one of two P-38Gs converted to night fighter specification by adding an SCR-540 radar unit and extended flash suppressors to the machine guns. The fighter is profiled with a detachable wing fuel tank sometimes carried on long-range operations. Note that a modified fuselage data stencil has been replicated in white stencils subsequent to the overall matt black paint job.

The subject of Profile 16 at Port Moresby's 12-Mile 'drome. A P-70 can be seen in the background.

One of the two P-38Gs operated from 12-Mile 'drome, Port Moresby, in 1943.

F-4 Lightning Limping Lizzie seen at 14-Mile. This view shows that the 8th PRS's drop tanks were also painted in the haze blue scheme.

The 8th PRS flight line at 14-Mile with a mixture of F-4s and F-5s.

CHAPTER 5
8th Photographic Reconnaissance Squadron

The 8th PRS was formed at March Field, California, on 1 February 1942 and comprised Flights A, B and C. As explained in Chapter 1, A Flight had arrived in Melbourne in April 1942 and had the distinction of being the first unit to operate Lightnings in the wider South Pacific region. However, B and C Flights did not arrive until July 1942, and it was not until then that the squadron began readying itself for the normal tempo of operations from its base at Townsville.

In the interim, the 8th PRS commander Lieutenant Karl Polifka, had maintained a largely solo effort that included a systematic mapping of enemy bases in New Guinea. These flights were staged through Port Moresby, and also surveyed potential locations for future Allied bases. As the sole photographic reconnaissance squadron in the SWPA the unit was administratively attached to the 19th BG for some months, before assignment to the 6th Photographic Reconnaissance Group.

The unarmed F-4 variant was fitted with two K-17 fixed vertical cameras in the nose compartment, configured for either vertical or oblique modes, exposing photos on large nine x nine-inch negatives. Each film magazine contained 250 exposures, and focal lenses of either six, 12 or 24 inches could be fitted depending on mission altitude, conducted between 10,000 to 30,000 feet. The highest mission altitude logged by the unit was at 37,000-feet to avoid a cumulonimbus frontal system. New Guinea's capricious weather proved more of a threat than Japanese fighters, and missions commonly lasted three and a half hours. The danger of flying an unarmed aircraft over enemy territory was over-stated. Even if intercepted at high altitude, a rare occurrence, a straightforward escape was to apply power and place the aircraft on a descent trajectory to outrun any pursuers.

Engine reliability was more problematic than first envisaged. Superchargers often failed, and contaminated fuel and magneto problems also contributed to downtimes and early returns. Other pernickety quirks to the oxygen system caused regular aborts.

By February 1944 the squadron boasted more than 30 pilots and moved to Nadzab in New Guinea's wide Markham Valley. The 8th PRS operated the F-4, F-5A, F-5B and F-5E models during its time in the SWPA. The squadron lost eighteen F-4s during its New Guinea operations; all were weather-related losses except for one shot down by a 13th *Sentai* Ki-45 twin-engine fighter over New Britain.

Markings

The F-4 was painted in a specialist blue speckled "haze" application which featured random layered shades of sky blue applied over a black base coat. Although considered effective at concealing the aircraft at high altitude, the paint was difficult to apply. The wear and tear of New Guinea's tropical climate soon desaturated the final layer exposing the black base primer in parts and in extreme weathering cases diminished the intended camouflage effect.

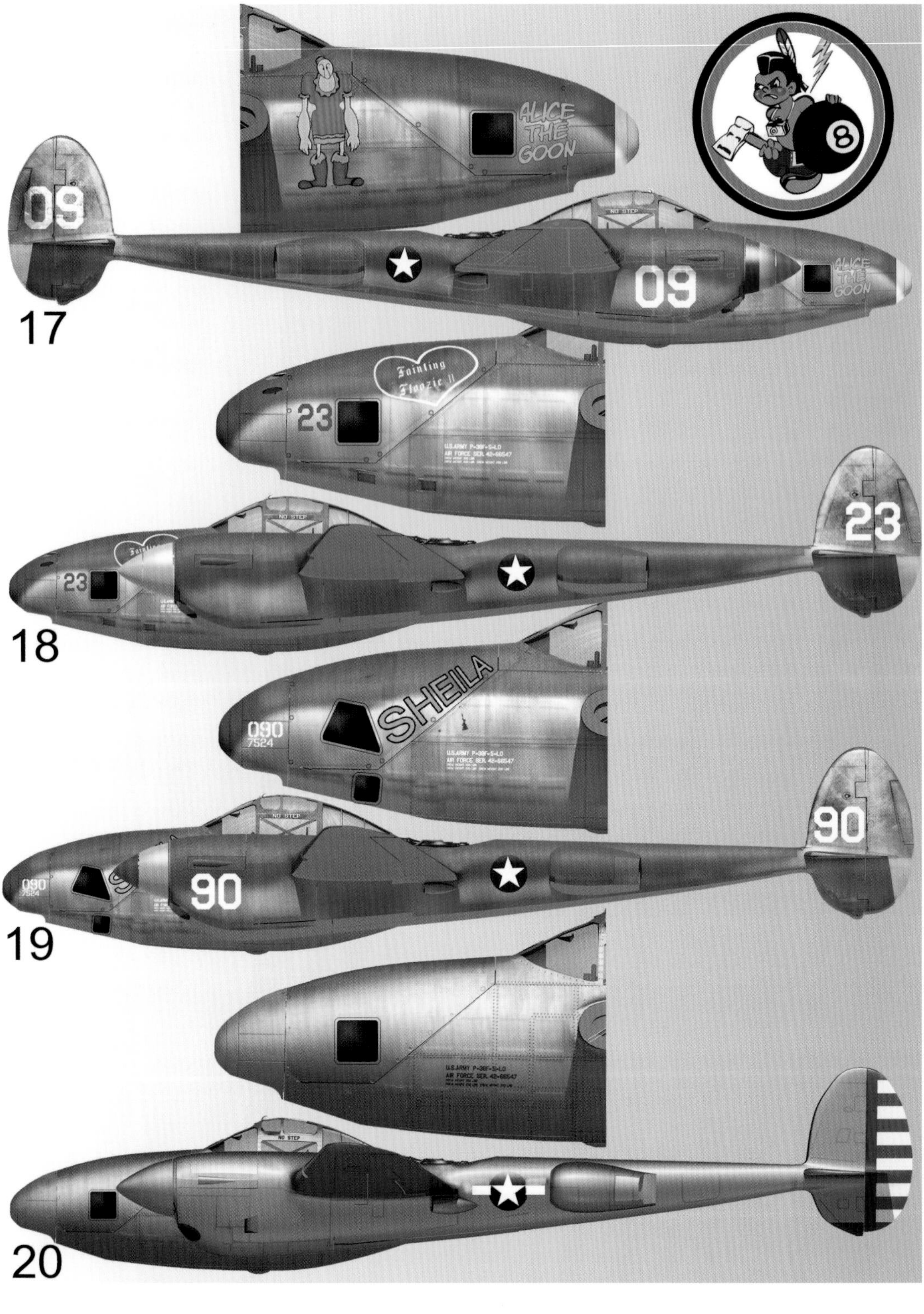
ALICE THE GOON
8
09
NO STEP
09
ALICE THE GOON
17
Fainting Floozie II
23
U.S.ARMY P-38F-5-LO
AIR FORCE SER. 42-66547
NO STEP
23
23
18
SHEILA
090
7524
NO STEP
090
7524
90
90
19
NO STEP
20

8th PRS Logo

The lightning bolt superimposed over clouds represents the Lockheed Lightning. The young Indian Brave represented the men in the squadron brandishing a hatchet to signify war.

Profile 17 – F-4 serial 41-2209, squadron #09, *Alice the Goon*

This F-4 was lost on 30 December 1943 when flown by Lieutenant Charles Erb. It was one of four 8th PRS reconnaissance Lightnings led by Lieutenant Walter Langdon which departed 14-Mile that morning at 0833. After photographing Gasmata and Arawe on New Britain, during the return journey Erb dropped out of formation and disappeared.

Profile 18 – F-4 serial 41-2123, squadron #23, *Fainting Floozie II*

This F-4 was named and often flown by the 8th PRS commander Lieutenant Karl Polifka. On 28 November 1942 just after take-off from 14-Mile 'drome, Lieutenant Philip Lissner feathered the starboard engine causing him to crash into a grass plain near 12-Mile 'drome, killing him instantly. The number 23 on the gondola was unusually painted in small size and in grey instead of white.

Profile 19 – F-5A serial 42-13090, squadron #90, *Sheila*

On 9 February 1944 this aircraft departed 17-Mile (Durand) flown by Flying Officer Andrew Davis and in formation with Lieutenant Orville Counselman flying F-5A serial 42-13093. The pair headed for Nadzab on a delivery flight but were forced to climb to 23,000 feet as there was cloud build-up over the ranges north of Nadzab. Counselman lost visual contact with Davis at 1410 when the pair turned back to find clearer weather. The wreckage of *Sheila* was located post-war in high mountains near Doli village, about 15 miles west of Salamaua.

Profile 20 – F-4 serial 41-2130, 14-Mile, October 1943

This F-4 was delivered to the Commonwealth Aircraft Corporation on 12 August 1942 for assembly. On 3 September it was flown to Townsville for assignment to the 8th PRS which named it *Malaria Mabel* with the squadron number 30. It was stripped down to bare natural metal finish for use as a squadron hack in October 1943, as illustrated here. On 3 November 1943 it was condemned by the Fifth Air Force however it was reassigned to the RAAF in February 1944. It is not clear from contradictory RAAF records whether it actually served briefly with the RAAF before being struck off charge by the USAAF in July 1944.

The subject of Profile 24, P-38J serial 42-104391, at Mono Island.

The subject of Profile 23, P-38J serial 42-104295, heads east along Guadalcanal's northern shore during a test flight around May 1944.

CHAPTER 6
12ᵗʰ Fighter Squadron "Dirty Dozen"

Constituted as the 12ᵗʰ Pursuit Squadron (Interceptor) on 20 November 1940, this unit was first assigned to the Seventh Air Force where it protected isolated Christmas Island during the bulk of 1942. In November of that year, it redeployed to Efate in the New Hebrides, before moving forward to Guadalcanal the following month.

At Guadalcanal the 12ᵗʰ FS operated from Fighter Strip Two, using P-39D, K and N model Airacobras to conduct escort missions, defensive patrols and later ground attack and bombing sorties. On 6 February 1943 the squadron moved to Treasury (Mono) Island in the north Solomons where it remained stationed until August 1944.

Several 12ᵗʰ FS pilots had been given opportunity to gain P-38 experience as early as April 1943 when they shared P-38s with the 339ᵗʰ FS, then operating P-38G and H models. The unit soon commenced operating its own P-38G and H models alongside Airacobras. Several of the squadron's pilots were included in the infamous 18 April 1943 "Yamamoto Mission". On 29 April 1943 the unit suffered its first Lightning loss when Lieutenant Gordon Whitaker was shot down in a P-38G over Bougainville escorting an F-5A on a photo-reconnaissance mission.

With Japanese air power all but gone from SOPAC skies by end of February 1944, the unit's Lightnings at Guadalcanal confined themselves to ground strikes as far as Rabaul, feeling peeved that the Fifth Air Force (referred to as "MacArthur's Air Force"), was getting far more publicity in the US.

The 12ᵗʰ FS pilots commenced rotations to Guadalcanal from late March 1944 to train on their new inventory of J model Lightnings, overseen by squadron commander Lieutenant Colonel Leland McGowan. In late 1944 the squadron moved to Morotai in the Netherlands East Indies before moving to the Philippines in early 1945.

The 12ᵗʰ FS lost nineteen Lightnings during its time in the SOPAC theatre: eleven to aerial combat, one to ground fire and seven to accidents or weather-related causes.

Markings

The squadron took delivery of a new inventory of P-38J model Lightnings in late March 1944 at Guadalcanal. These were J-15 models followed a few weeks later by an additional batch of similar J-20s. These natural metal finish models received squadron numbers commencing in the 200 to 229 range, always applied to the nose, and sometimes also to the air scoops and/or fin.

This numbering system was a follow-on from late 1943 when the Thirteenth Air Force instituted a three-digit system whereby Airacobra units could be identified in the air: 200 series for the 12ᵗʰ FS, 300 series for the 70ᵗʰ FS and 400 series for the 68ᵗʰ FS. When the system was changed again to accommodate different aircraft types, the 12ᵗʰ FS shared the 200 series with 68ᵗʰ FS P-40Fs.

Profile 21 – P-38H serial 42-66683, squadron #208, *Pluto*, Munda, January 1944

P-38H *Pluto* was one of 27 P-38H models which served in the 12th FS inventory in the serial range of 42-66583 to 42-66899. It had one Japanese flag painted over the yellow Constructor's Number stencil 1194. The aircraft was written off following an operational accident in February 1944. The name *Betty* appeared on the port engine cowl, applied by the crew chief.

Profile 22 – P-38G serial 43-2218, squadron #18, Munda, April 1943

The squadron number 18 remained extant on this airframe after it was transferred from an unidentified unit. After departing Fighter Two on Guadalcanal, on 29 April 1943 Lieutenant Gordon Whitaker and squadron commander Major Louis Kittel escorted a 17th PRS F-5A to Bougainville. Whitaker experienced engine trouble and had trouble keeping up with the F-5A. He fell behind and around mid-morning Whitaker broadcast he was baling out over Bougainville's west coast, which was enemy held territory. He had been attacked by a formation of six No. 582 *Ku* Zeros based at Buin led by Warrant Officer Tsunoda Kazuo who recorded that it went down in flames. Whitaker remains Missing in Action.

The subject of Profile 21, P-38H serial 42-66683, undergoing maintenance on Guadalcanal.

P-38J-20 serial 44-23291 being assembled on Guadalcanal. The digits 4295 are, exceptionally, the Constructor's Number, and not the last four digits of the serial number. The smaller stencilled numbers are factory batch and order numbers.

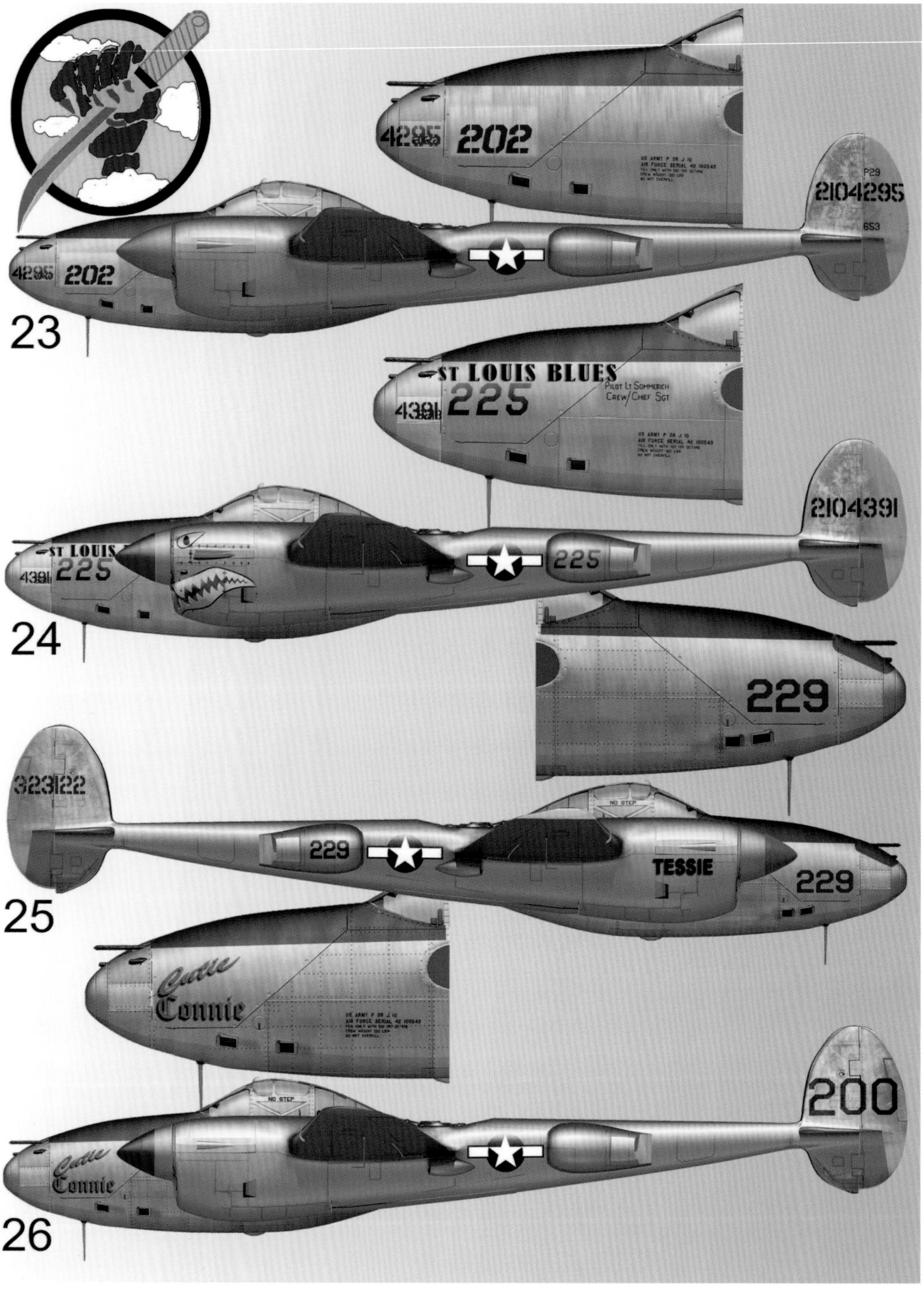
4295
202
2104295
653
23
ST LOUIS BLUES
225
4391
PILOT LT. SOMMERICH
CREW/CHIEF SGT
ST LOUIS
2104391
24
229
323122
TESSIE
NO STEP
25
Cutie
Connie
200
26

12th FS Logo– A clenched fist holding a knife against a yellow sky with white clouds.

Profile 23 – P-38J serial 42-104295, squadron #202, Guadalcanal, March 1944

This fighter is illustrated as a brand-new airframe, unpacked and assembled at Guadalcanal. It was the third P-38J-15 assembled by the 12th FS and was thus allocated squadron number 202, after the first and second Lightnings were allocated 200 and 201 respectively. The last four digits of the serial were stencilled over the Constructor's Number 3122. The small stencils P29 and 653 on the rudder were applied by a sub-contractor who assembled the rudders separately from the main Lockheed factory.

Profile 24 – P-38J serial 42-104391, squadron #225, *St Louis Blues*, Guadalcanal, May 1944

Following assembly at Guadalcanal in late March 1944, this Lightning was assigned to Lieutenant Eugene Somerich who named it after a popular song of the time, *St Louis Blues*. First published in 1914, the song later prospered as a pillar of the US jazz repertoire often played by the Benny Goodman Orchestra throughout the 1940s. The last four digits of the serial number are stencilled over the Constructor's Number 3218. The shark's teeth were added just before departure for Morotai and the boom airscoops were later decorated in blue checkers in the Philippines.

Profile 25 – P-38J serial 43-23122, squadron #229, *Tessie*, Mono Island, July 1944

This Lightning flew several ground strikes in the theatre before departing for Morotai. It only carried the name on the cowl and was lost on 11 November 1944 during a flight from Morotai by Lieutenant Robert Russell. Note the white wing tips painted as a formation marking.

Profile 26 – P-38J 44-23287, squadron #200, *Cutie Connie*, Mono, July 1944

This Lightning flew only briefly in the South Pacific, conducting several ground strikes against New Britain from Mono Island before departing for Morotai.

The subject of Profile 25, P-38J serial 43-23122, at Guadalcanal shortly after assembly. The starboard prop carries 3122-R in chalk, showing it has just been freshly attached from the packing case.

The subject of Profile 27, F-5A serial 42-12670, at Guadalcanal's Fighter Two strip which had been bulldozed from a Lever Brothers coconut plantation.

CHAPTER 7
17th Photographic Reconnaissance Squadron

The 17th Photographic Reconnaissance Squadron was created and assigned to the 4th PRG on 14 July 1942. The squadron was initially equipped with the F-5A reconnaissance version of the Lightning, and later also operated F-5Bs and F-5Es.

The air echelon was assembled at Colorado Springs, Colorado, and commanded by Major John Murray. On 2 December 1942 the headquarters squadron along with A and B Flights unloaded their F-5As at Noumea which had been shipped across the Pacific as deck cargo. Meanwhile C and D Flights remained in the US until January 1944. Following a month of aircraft assembly and test flights around New Caledonia, the squadron flew its first combat mission from Guadalcanal's Fighter Two on 5 February 1943. This airfield, freshly graded and surrounded by high palm trees, became home to the squadron which commenced operations with eight aircraft on charge. It flew daily missions, and chalked on the bottom of its operations blackboard was an order to pilots that the only identification they should carry in the cockpit were dog tags. The only weapons carried by pilots on long-range missions were a 0.45-inch calibre pistol and a machete.

The 17th PRS lost its first Lightning on 14 February 1943 when Lieutenant Ardell Nord went Missing in Action on a reconnaissance mission over Kahili Airfield in southern Bougainville. Japanese records show that he had been shot down by four No. 252 *Ku* Zeros. Another substantive loss was on 28 August 1944 when Lieutenant Stanley Alexander's F-5E disappeared on an altitude test flight from Guadalcanal. A USN destroyer near the Russell Islands saw an aircraft spin and crash into the water. Parts of the retrieved debris were recognised as belonging to Alexander's F-5E.

The 17th PRS lost six aircraft during combat missions and four to accidents in the theater. It finished the war operating from Palawan in the Philippines.

Markings

The 17th PRS stencilled the last three digits of the serial number on the nose as a squadron number. It received its first F-5Bs in March 1944 which arrived in a revised factory finish of overall dark blue, a scheme which commenced with the first batch of F-5A-3-LOs.

17th PRS Logo

"Heckle" the Walt Disney woodpecker, standing on a cloud and holding a camera.

Profile 27 – F-5A serial 42-12670, squadron #670, Guadalcanal, May 1943

This converted P-38F had the Constructor's Number 7104 stencilled on the nose over-painted with the last three digits of the serial number. It was not lost to operations or combat while with the 17th PRS.

Profile 28 – F-5A serial 42-12972, squadron #972, Guadalcanal, September 1943

This F-5A-3 airframe was painted overall dark blue, with the US "stars and bars" insignia, denoting a change from the two-tone colour scheme of the F-5A-1 as illustrated above. Unit maintenance logs show it had accumulated 420 hours flying by the time it went missing on 6 December 1943 when flown by Lieutenant James Reed. It had departed Munda at 1505 that afternoon for a three-hour reconnaissance of southern Bougainville. No Japanese fighter units report engaging a P-38 over Bougainville that afternoon, so it is likely a weather-related loss.

The subject of Profile 28, F-5A serial 42-12972, being assembled at Fighter Two around late September 1943.

The camera arrangement in the F-5 series.

The subject of Profile 29, F-5B serial 42-67354, at Nadzab.

*The 25th PRS Operations Officer, Captain Thomas Robertson, with
F-5B serial 42-67357 Paper Doll at Nadzab around June 1944.*

CHAPTER 8
25th Photo Reconnaissance Squadron "Hawkeyes"

Both this squadron and the 26th PRS (see Chapter 10) were formed at Peterson Field, Colorado, on 9 February 1943. They were assigned to the 6th PRG, which also incorporated the 8th PRS, already operational in the SWPA. Both the 25th and 26th PRS crossed the Pacific aboard the liner *Niue Amsterdam* before arriving in Sydney on 19 November 1943. The 25th PRS then proceeded by rail and reformed in Brisbane the following week. By January 1944 it had made its way north to Port Moresby, with the move overseen by Operations Officer Captain Thomas Robertson.

Meanwhile, several 25th PRS pilots had already arrived in New Guinea in early December where they spent several weeks at Schwimmer 'drome (14-Mile) attached to the 8th PRS for familiarisation purposes. During this period Flying Officer Fred Cross was killed in a training accident on 5 December 1943, becoming the unit's first overseas loss.

After receiving its own aircraft, the 25th PRS moved to Nadzab on 7 February 1944. After five months the unit started packing up at Nadzab for a move to Biak in Dutch New Guinea on 23 July. In the SWPA the 25th PRS accepted tasking from both the Fifth Air Force, and/or the US Sixth Army via a US Army liaison officer assigned with the unit. During its time in the SWPA, the squadron lost two aircraft to operational causes and three went missing.

Markings

The 25th PRS initially borrowed 8th PRS F-4s to conduct familiarisation training in Port Moresby. It later operated the F-5A and F-5B in the SWPA, then later F-5Es when it moved to the Philippines. Its initial batch of F-5As had been assembled at Amberley and these were assigned as new aircraft. As the campaign wore on there were often airframe swaps between all three 6th PRG reconnaissance squadrons, effected mostly by maintenance and repair cycles. The unit started paint-stripping its early model F-5As around the end of March 1944 at Nadzab, which gave them an extra 10 miles per hour cruise speed. New F-5Bs were then being delivered in natural metal finish.

Squadron numbers were applied as the last three digits of the serial on the nose. Propeller spinners were prominently marked to facilitate aerial identification using varying colours, usually involving red. This habit developed into bands on the spinners, and later when the unit moved to the Philippines the tails of its Lightnings were painted with the diving eagle motif of the 6th PRG. The 25th PRS carried the fewest examples of nose art out of the 6th PRG's three squadrons.

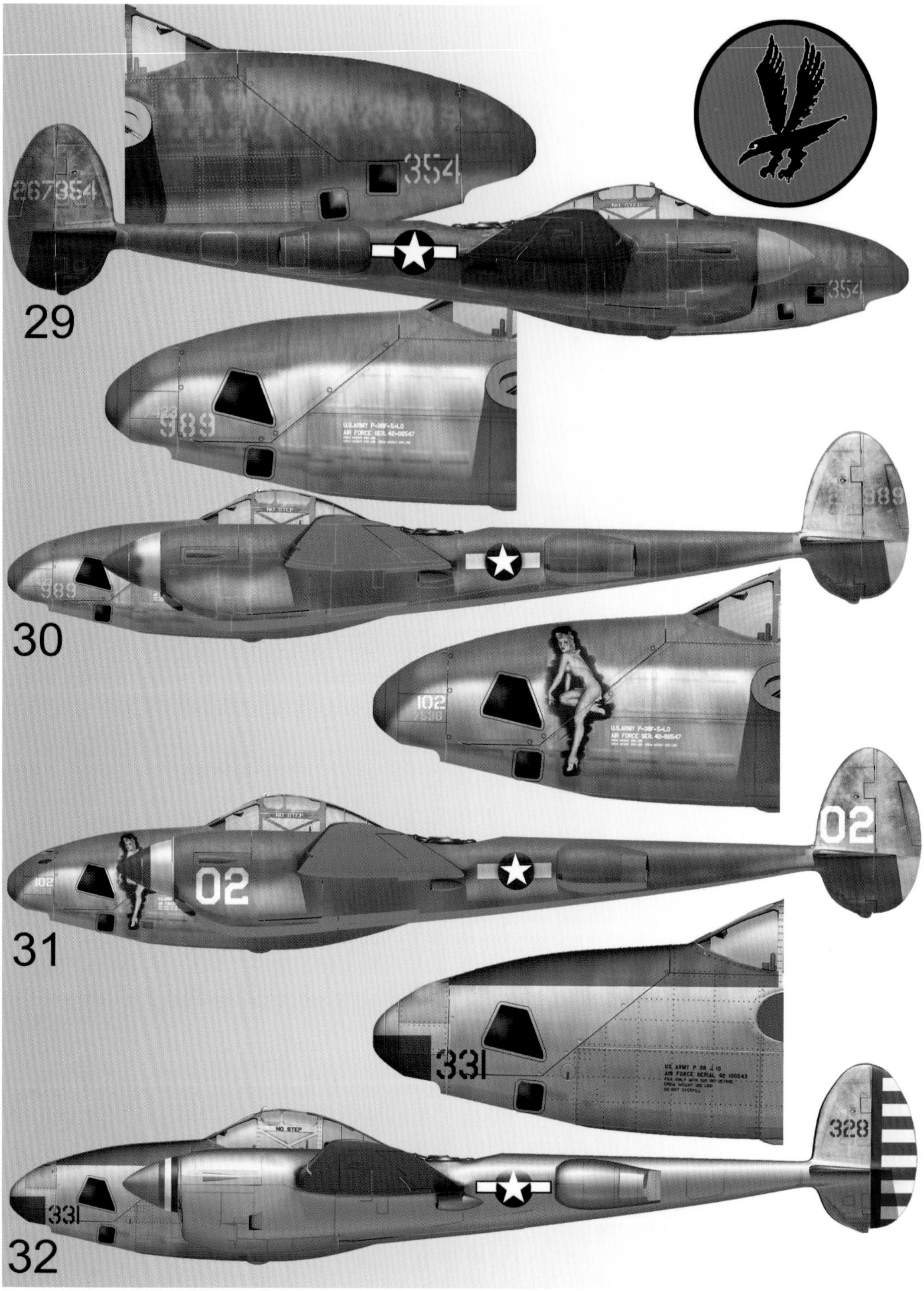
29
267354
354
NO STEP
30
989
NO STEP
31
102
02
NO STEP
32
331
328
NO STEP

25th PRS Logo

A black diving eagle over a green background.

Profile 29 – F-5B serial 42-67354, squadron #354, Nadzab, April 1944

This converted J model airframe was painted in the later overall darker blue haze scheme and had yellow spinners. It went missing on 3 April 1944 when returning to Nadzab from a strike against Hollandia. The last three digits of the serial number were stencilled on the nose as a squadron number.

Profile 30 – F-5A serial 42-12989, squadron #989

The last three digits of the serial number were used as the squadron number. The forward section of the spinners were sprayed in blue.

Profile 31 – F-5A serial 42-13102, squadron #02

This aircraft was the mount of the commander of the 6th PRG, Colonel Ben Armstrong. Originally maintained by the 8th PRS it was regarded as a headquarters aircraft and was often flown by 25th PRS pilots at Nadzab. It has the early haze scheme as applied to the first batches of F-5As with added white bars to the star. The aircraft carried no name, and the art on the gondola was copied directly from a Vargas calendar. The last two digits of the serial were painted on the fin and engine by the 8th PRS. The small white stencil 102 was applied over the Constructor's Number 7536 which was factory stencilled in yellow. The aircraft went missing on 2 July 1944 after departing Nadzab. It was being flown by 8th PRS pilot Lieutenant Thaddeus Jones to obtain low-level oblique photographs of Wewak's airfields. Over the target Jones and an accompanying F-5A drew anti-aircraft fire from Wewak Point, and Jones failed to return.

Profile 32 – F-5B serial 43-28331 (the last three digits painted over), squadron #331

This airframe was delivered in natural metal finish in late March 1944 and operated from Nadzab before the unit later transferred to Biak. The lower section of the nose was painted in matt black, with decorated spinners and rudders.

This frontal view shows the F-5's twin drop-tanks which boosted its range considerably.

The 9ᵗʰ FS flight line at Dobodura. Squadron number 88's crew chief looks at the camera, while squadron number 95 has a flight leader band on the boom.

CHAPTER 9
9th Fighter Squadron "Flying Knights"

One of the 49th FG's three squadrons, the 9th FS had spent most of 1942 defending the Darwin area with P-40Es. After leaving that theatre around late September, it was intended to convert to the new Lightnings for use in New Guinea. However, the 9th FS's association with the P-38 was at times convoluted and not without rancor.

When the first batch of sixteen P-38Fs earmarked for the squadron arrived in Townsville in October 1942, they were assigned to a contingent of 17th Fighter Squadron (Provisional) pilots, with previous combat experience in Java. These pilots were posted into the 9th FS, and the squadron commander Jesse Peaselee appealed to Fifth Fighter Command to find P-38 assignments for the squadron originals who felt left out. However, the "originals" languished in Townsville throughout October pending a decision.

Meanwhile, the sixteen P-38Fs did not meet the quantity needed by the squadron to be fully operational.[1] New P-38Gs were becoming available, but technical problems had complicated their introduction into service. Attempts were made to rectify these problems with mixed success. For example, at end of October five P-38Gs were flown to 30-Mile where leaking rubber fuel tanks were replaced with new ones. However, other technical challenges remained which mirrored the situation in the 39th FS (which had been operational since September – see Chapter 3), and the 9th FS's G-models did not enter frontline service for another six weeks.

As an interim measure and to stop his best pilots from applying for combat assignments with other units, Peaslee organised a group to serve a secondment with the 39th FS to gain frontline Lightning experience. It was not until January 1943 when the 9th FS finally had a full complement of its own P-38s, and the unit began using them for familiarisation flights. New Guinea's mountainous geography was still largely an unknown, especially since the Darwin area had a comparatively flat landscape. Indeed, there was considerable embarrassment when on a 20 February 1943 escort mission to Gasmata, the 9th FS force-landed three P-38Gs on the return journey. Lieutenants Arthur Bauhof, Robert Douglas and Harry Lidstrom all put down on a beach near a plantation at Kerema, some 140 miles northwest of their departure point. Whilst all three airframes were later disassembled and returned by barge, the event attracted unwanted mockery around Port Moresby's airfields.

Lockheed technician Frank Bertelli resided with the 9th FS during this early phase, advising how best to keep the big fighters airborne and assisting with field modifications to improve performance and combat effectiveness. On 6 March 1943 the squadron moved to Dobodura, and in late 1943 staged from Kiriwina in the Trobriand Islands for a series of strikes against Rabaul. In December 1943, due to ongoing shortages of Lightnings in the Fifth Air Force and

---

1   At this time Fifth Air Force P-38 squadrons generally needed 25 aircraft to be at full strength, supported by a reserve pool used for attrition replacements and maintenance rotations.

with an element of bitterness, all of the squadron's P-38s were transferred to the 475th FG. The 9th FS then transitioned to the P-47D Thunderbolt which it operated from Gusap until May 1944, after which it transferred out of the SWPA to Dutch New Guinea and then the Philippines.

During its first SWPA tour with Lightnings from January to December 1943 the squadron lost eleven Lightnings to combat, of which five were missing. Extraordinarily, it lost twice as many airframes to other non-combat causes, exactly 22.

Markings

The 9th FS used unit numbers in the 70 to 99 range, the same as they had on their previous P-40s, applied to both sides of the nose and tail.

Squadron number 79 was P-38H serial 42-66847 assigned to Lieutenant Richard Bong, hence explaining the sixteen kill markings. This photo was taken at Dobodura subsequent to a landing accident sometime after 28 June 1943, the date of Bong's sixteenth claim. This Lightning was sent for repairs and was replaced in 9th FS service by another squadron number 79.

P-38G squadron number 75 at Dobodura following a rainy evening. Note the canvas covers over the superchargers.

P-38G squadron number 96 following a landing accident at Dobodura.

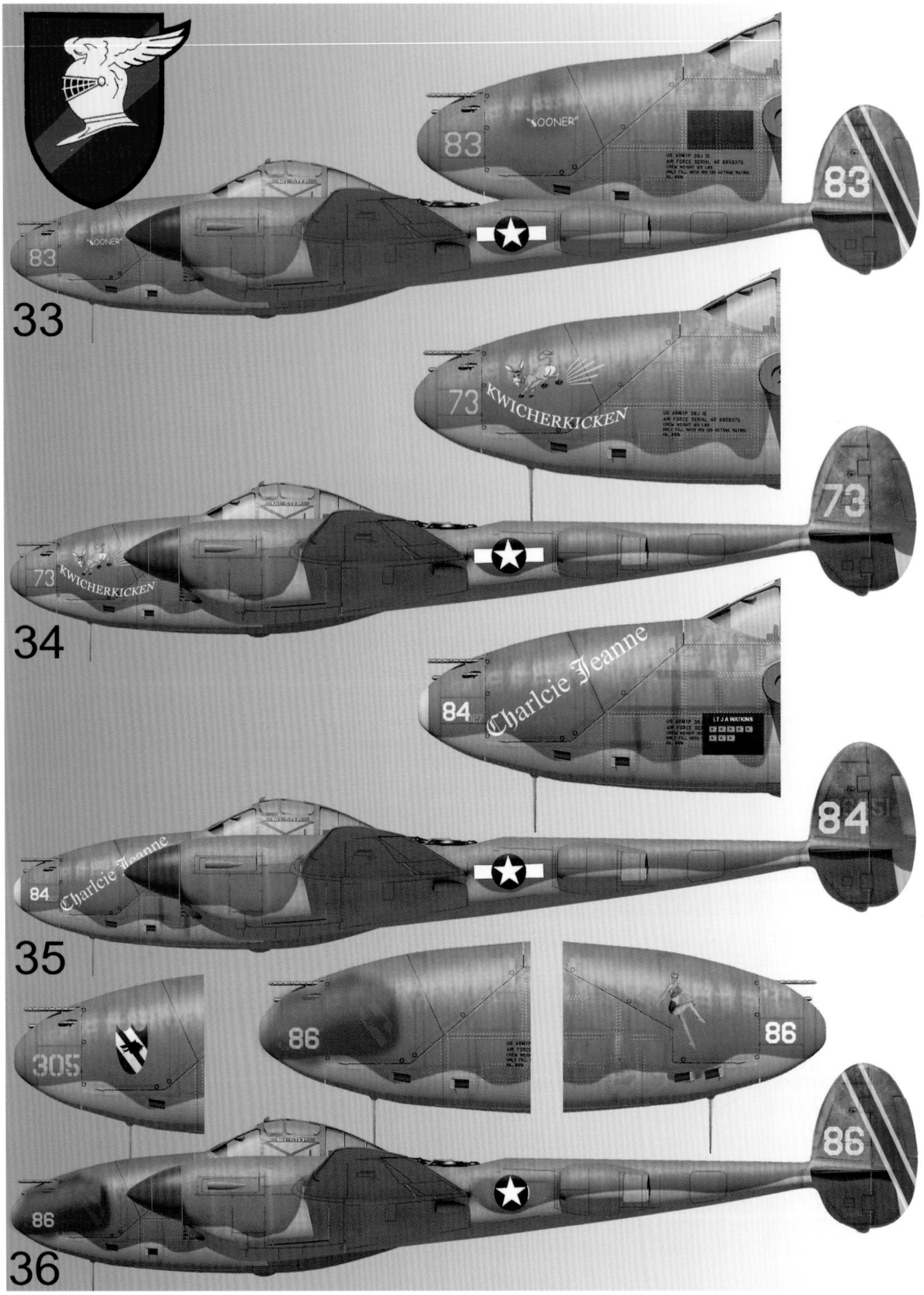
"SOONER"
83
33
KWICHERKICKEN
73
34
Charlcie Jeanne
84
LT J A WATKINS
35
305
86
36

9th FS Logo

The "Flying Knights" logo was a knight's head armour with bird wings attached.

Profile 33 – P-38F serial 42-12655, squadron #83, *Sooner*

This fighter was assigned to Captain Gerard Johnson who immediately arranged to change the squadron number to 83 since this was his lucky football jersey number at Oregon High School. On 5 November 1943 Lieutenant George Haniotis went missing in this aircraft during extensive combat over Rabaul. The 9th FS had staged through Kiriwina airfield and on this mission, escorting B-24 Liberators, Haniotis was part of Red Flight flying as wingman for flight leader Captain Richard Bong. While over Tobera airfield (southeast of Rabaul), Haniotis called Bong over the radio twice and asked if he had left the area. This last transmission was made at 1240, and although the reason for his loss is unknown, it is likely he was shot down by Zeros. Johnson was later appointed commander of the 9th FS when it converted to P-47Ds.

Profile 34 – P-38G serial unknown, squadron #73, *Kwicherkicken*

This fighter appears in several photos taken at Dobodura, however little is known of its history, except that it was assigned to and named by Lieutenant James Haislip.

Profile 35 – P-38H serial 42-66516, squadron #84, *Charlcie Jeanne*

Named by and assigned to Lieutenant James Watkins, on 13 October 1943 whilst inbound to Rabaul on an escort mission, the 9th FS encountered marginal visibility and rain. Three Lightnings failed to exit the bad weather, and Captain Gerald Johnson who was leading the flight later submitted that the pilots were likely lost to a mid-air collision as each had sufficient instrument experience to safely exit the weather. One of the missing Lightnings was *Charlcie Jeanne* flown by Lieutenant Theron Price.

Profile 36 – P-38G serial 43-2196, squadron #86

This fighter was assigned to squadron leader Sidney Woods in late 1943. The fighter was one of several Guadalcanal veterans which had served with the 44th FS prior to refurbishment in Townsville. The digits 305 are the last three digits of the Constructor's Number 3305 which was applied in the Townsville overhaul depot. The meaning of the previous badge with a wolf's head, later painted over when the fighter was reassigned to Woods, is unclear. The girl on the starboard nose was a copy of an Alberto Vargas calendar girl. The fighter was lost on 10 October 1943 when Lieutenant Clayton Barnes crashed on take-off at Dobodura following an engine failure.

Profile 37 – P-38G serial 43-2204, squadron #99, *Beautiful Lass*

Assigned to and named by Lieutenant Jerry O'Neill, the *Beautiful Lass* artwork was a direct copy of an Alberto Vargas calendar girl. Sometime after claiming his eighth victory, the nose art was repainted (as illustrated here), making the female white-skinned and repainting the name *Beautiful Lass* in white letters. The black and white scoreboard denotes eight combat claims O'Neill made between March and October 1943. After O'Neill's return to the US the fighter was overhauled by a service squadron and reassigned to the 431st FS. It went missing with this unit on 28 December 1943 when flown by Lieutenant Ormond Powell. During a patrol over Cape Gloucester the weather closed in, and the formation climbed to 25,000 feet to get over it, but Powell lagged. The flight leader Lieutenant Paul Morris instructed him not to enter the overcast, however Powell was seen to enter the front. When the others broke clear, Powell had disappeared, and he remains Missing in Action to this day.

Profile 38 – P-38F serial 42-12652, squadron #82, *Double Trouble*

This fighter started its career as squadron number 33 with the 39th FS and also served briefly with the 80th FS from late February 1943. When assigned into the 9th FS it was given the squadron number 82 and was named by its assigned pilot Lieutenant Frank Wunder. It was later renumbered 91 following repairs, before being assigned to the 431st FS, meaning that it thus flew with every Fifth Air Force Lightning fighter group. It is currently under restoration in California.

The nose art seen on Profile 37 at Dobodura.

Refuelling squadron number 74 at Dobodura.

An F-5B undergoes an engine change at Nadzab in May 1944.

Captain John Carner with the subject of Profile 39, F-5B serial 42-68248.

CHAPTER 10
26ᵗʰ Photographic Reconnaissance Squadron

Both the 25ᵗʰ PRS (see Chapter 8) and the 26ᵗʰ PRS were formed at Peterson Field, Colorado, on 9 February 1943 and assigned to the 6ᵗʰ Photographic Reconnaissance Group. Both squadrons crossed the Pacific together aboard the liner *Nieu Amsterdam*, which transited Wellington, New Zealand, on its way to Sydney where it docked on 19 November 1943.

The 26ᵗʰ PRS's ground echelon proceeded north to Brisbane the following week before shipment direct to Dobodura in January 1944. After several weeks of operations at Dobodura, the squadron departed for Finschhafen on 18 February 1944. Meanwhile the air echelon commenced its first operations by borrowing F-5As from the 8ᵗʰ PRS at Port Moresby before moving to Finschhafen and then Nadzab on 29 March.

The Operations Officer Captain Sheldon Hallett and squadron commander Major Walter Hardee, who later named his F-5B *Anne II*, brought the unit to New Guinea. Hallett became the unit's first combat loss when he went missing in an F-5B on a mission to photograph Tadji on 29 February 1944. In June 1944 the unit moved out of the SWPA theatre when it departed Nadzab for Dutch New Guinea. During its time in the SWPA, and similar to the 25ᵗʰ PRS, it routinely accepted tasking from both the Fifth Air Force and also direct from the US Sixth Army via a US Army liaison officer assigned to the unit.

Markings

The 26ᵗʰ PRS mostly operated the F-5B in the SWPA, and later the occasional F-5E, both of which were delivered in natural metal finish. In the last few weeks at Nadzab it took delivery of several F-5Es which it took to the Philippines. Its initial batch of F-5Bs was assembled at Townsville and assigned as new aircraft. As the campaign wore on, many airframes transited between all three 6ᵗʰ PRG reconnaissance squadrons, caused mostly by maintenance and repair cycles. Similar to its sister squadron, the 25ᵗʰ PRS, the squadron's propeller spinners were prominently marked to facilitate aerial identification using varying colours, usually involving red. Corporal Wichman developed expertise in painting precision spinners and became the *de facto* 26ᵗʰ PRS artist. The squadron used the last three digits of the serial number, as stencilled in the factory on the nose, as the squadron identifier.

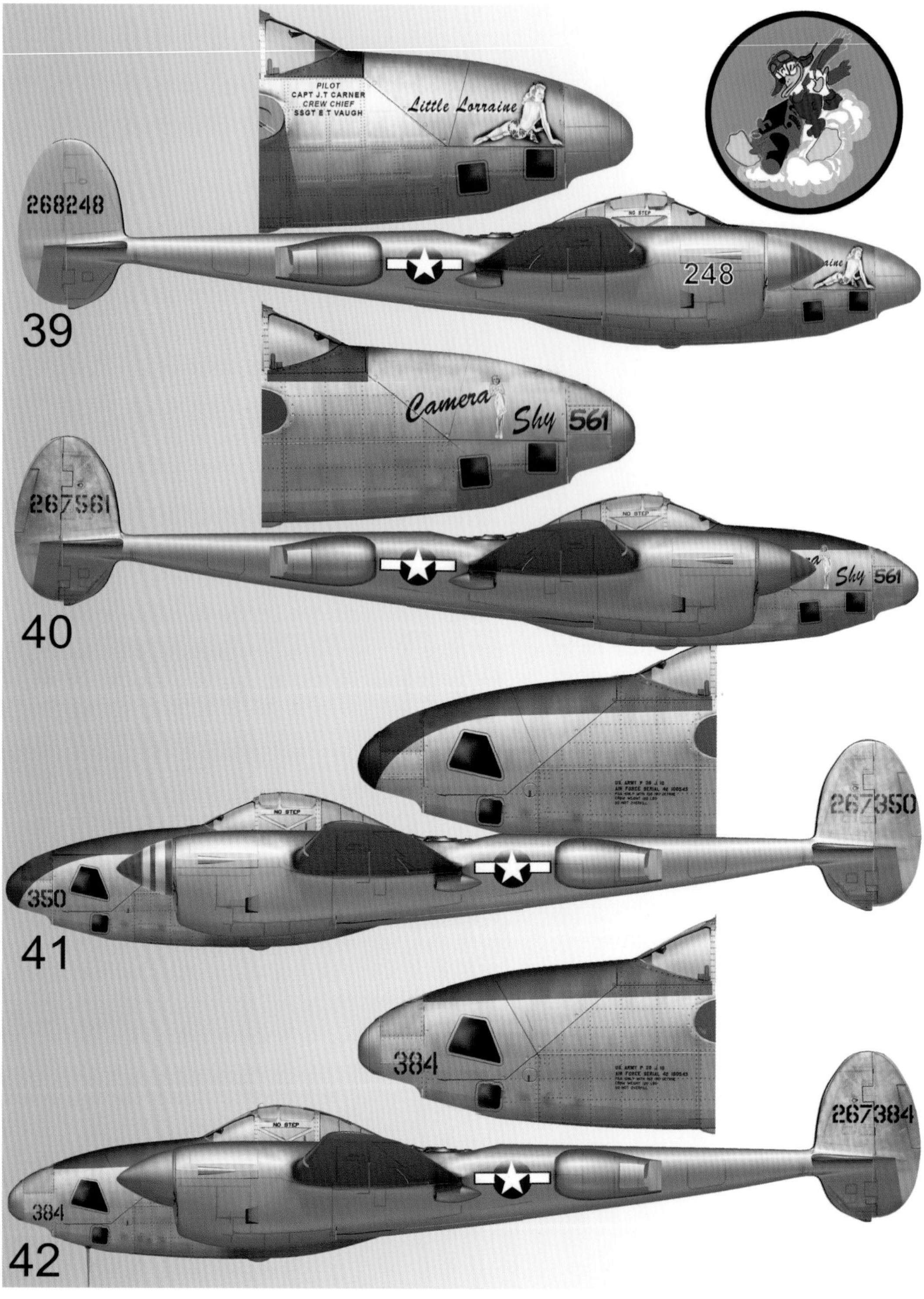
PILOT
CAPT J.T CARNER
CREW CHIEF
SSGT E.T VAUGH
Little Lorraine
268248
NO STEP
248
39
Camera
Shy
561
267561
NO STEP
561
40
267350
NO STEP
350
41
384
267384
NO STEP
384
42

26th PRS Logo

Donald Duck, sitting on a cloud with one eye squinted and holding a large aerial camera.

Profile 39 – F-5B serial 42-68248, squadron #248, *Little Lorraine*

This Lightning was among the first batch assigned to the 26th PRS in late January 1944 and was later assigned to and named by Captain John Carner. He was appointed Operations Officer to replace Captain Sheldon Hallett who went missing on 29 February 1944. When the unit started moving out of Nadzab in preparation for its move to Dutch New Guinea, many of the original F-5Bs were transferred to the 8th and 25th PRS including *Little Lorraine*, which was transferred to the former squadron on 23 June 1944.

Profile 40 – F-5B serial 42-67561, squadron #561, *Camera Shy*

This F-5B was transferred from the 26th PRS to the 25th PRS on 28 July 1944 at Nadzab. Note the squadron number 561 has been applied in a stylised calligraphy and not in stencil format.

Profile 41 – F-5B serial 42-67350, squadron #350

On the infamous "Black Sunday" mission of 16 April 1944, this Lightning (callsign "Topmost") was flown by Lieutenant Donald Christians who let down in Saidor's circuit area very low on fuel. After landing he collided head-on with a 405th BS Mitchell about halfway down the runway. The resultant wreckage of both aircraft burned fiercely, rendering the strip unusable until it was cleared. Christians was killed immediately upon impact. Note the striped spinners and the extended anti-glare panel on the gondola.

Profile 42 – F-5B serial 42-67384, squadron #384

The 26th PRS Operations Officer Captain Sheldon Hallett was flying this unnamed Lightning when he was shot down near Tadji by JAAF fighters on 29 February 1944. Note that the squadron number has been applied over the faded Constructor's Number 1895.

Brown Eyes was an F-5B (serial unknown), seen at Nadzab in June 1944.

P-38L serial 44-26306 was assigned to 35th FS commander Major Harold McClelland. Taken later in the war, this Lightning showcases a polished rear section spinner with a yellow forward half delineated by a black stripe, a practice which commenced at Nadzab with the squadron's natural metal finish P-38s.

The art applied by Captain Harley Brown of the 431st FS on P-38H serial 42-66746 as per Profile 45. The Latin motto of "Nemo Me Impune Laccessit" translates as "No-one provokes me without impunity".

CHAPTER 11
35th Fighter Squadron "Black Panthers"

The three 8th Fighter Group squadrons, the 35th, 36th and 80th FS, arrived in Australia in late February 1942, equipped with Airacobras. In subsequent months of that year all three squadrons saw much combat in New Guinea, from bases at Port Moresby and Milne Bay.

The 35th FS enjoyed a major break from combat at the end of 1942, before returning to New Guinea in January 1943. Due to the shortage of P-38s they continued using Airacobras, with the squadron's last P-39 loss occurring near Port Moresby on 27 July 1943. From October 1943 the unit transitioned to P-40N Warhawks, while a shortage of Lightnings saw it briefly operate P-47D Thunderbolts.

The first Lightnings were finally delivered to the 35th FS at Nadzab in late March 1944, and the squadron operated the type until the end of the war. Major Harold McClelland was appointed squadron commander in January 1944, and he led the unit throughout its P-38 era in the SWPA. However, this era was brief, as by July 1944 the squadron had moved to Dutch New Guinea.

While in the SWPA, the 35th FS had three Lightnings downed by Japanese aircraft. Another four went missing, while thirteen others were lost to various other causes including air raids.

Markings

From the commencement of its time in theatre the 35th FS marked its fighters with alphabetical identifiers. It used the squadron colour of yellow on its P-38s, and this marking was often accompanied by black piping to delineate the colour. Squadron letters were often "shadowed" for the same purpose. The squadron was initially given a large batch of former 475th FG P-38Fs to commence operations, followed by batches of new P-38Js. By war's end it was operating mostly P-38Ls.

The subject of Profile 45, P-38H serial 42-66746, on the 431st FS flight line at Nadzab, just before it was transferred to the 35th FS.

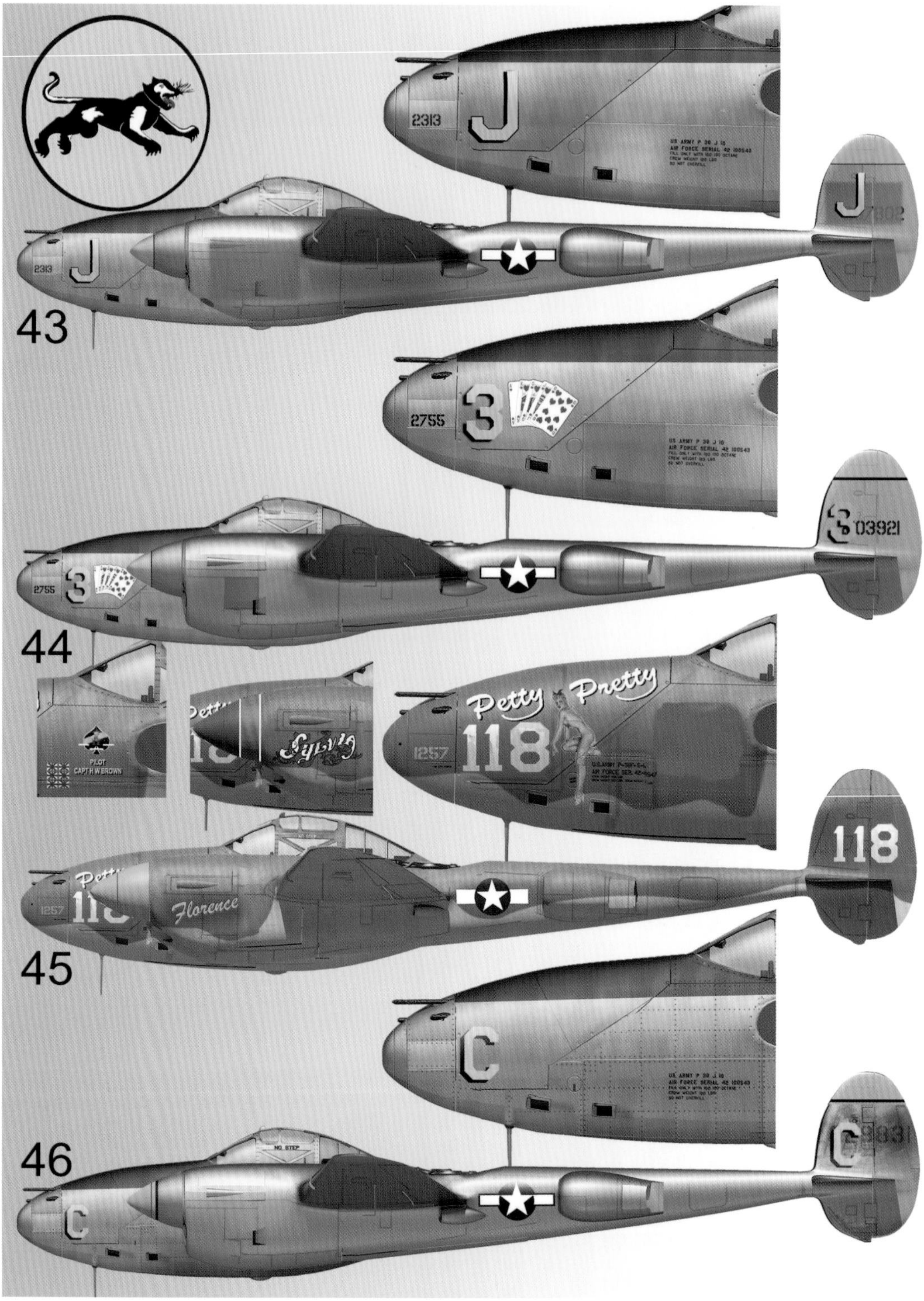
2313
J
43
2755
3
03921
44
PILOT
CAPT H W BROWN
Sylvia
Petty Pretty
1257
118
Florence
45
C
NO STEP
46

35ᵗʰ FS Logo

A running black panther with blood in its mouth.

Profile 43 – P-38J serial 42-67802, squadron letter J, Nadzab, May 1944

This P-38J-5 previously served with the 431st FS. It was transferred to the 35ᵗʰ FS after undercarriage repairs, during which it was stripped down to natural metal finish, thus removing all previous 431ˢᵗ FS markings. The letter J was applied on the nose and the overhaul depot re-stencilled the Constructor's Number 2313 on the nose. On 18 June 1944 Lieutenant Delmar King was strafing Dagua airfield after which he broadcast that he had been hit in the left engine and dropped back. With fuel streaming from the damaged engine, King bailed out about a mile off the Murik Lakes and was seen floating with his Mae West and gathering his parachute. A patrolling VP-34 PBY landed nearby to collect him but failed to find King who remains Missing in Action.

Profile 44 – P-38J serial 42-103921, squadron #3, Nadzab, April 1944

Lieutenant Richard West mostly used this fighter which was delivered to the 35ᵗʰ FS at Nadzab in April 1944. The nose art is a flush of hearts, ten through to ace, and the black Constructor's Number stencil 2755 was visible on the nose. For an unknown reason the numeral 3 was applied instead of an alphabetical squadron identifier. When the numeral was applied on the tail too, the black serial stencil remained visible. This Lightning was lost during a Japanese air raid at Wakde Island on 6 June 1944.

Profile 45 – P-38H serial 42-66746, squadron #118, *Petty Pretty*, Nadzab, March 1944

This Lightning went missing on 30 March 1944 piloted by Lieutenant Dean Jacobsen who had departed Nadzab for an administrative flight to Milne Bay. The aircraft had just been reassigned from the 431ˢᵗ FS and still retained that squadron's markings as named in that unit by Captain Harley Brown before it was taken over by Lieutenant Frank Monk. The crew chief at the time had named the left engine *Sylvia*. Brown's previous markings are also showcased. After arrival at the 35ᵗʰ FS, the engines were subsequently renamed by the ground crew *Doris* and *Florence*, and the spinners and tips of the tailplane were painted the squadron colour yellow.

Profile 46 – P-38J serial 43-28831, squadron letter C, Nadzab, March 1944

This P-38J was assigned to Lieutenant Clifton Troxell and maintained by crew chief Sergeant Peter Gino. Troxell offered the Lightning to Charles Lindberg to fly on his second familiarisation flight from Nadzab on 16 June 1944. During the uneventful flight Lindberg flew as wingman to Troxell where they circled shipping in Huon Gulf before returning to Nadzab.

The subject of Profile 48 at Nadzab shortly after it was delivered and decorated.

Lieutenant Erdmann with his hand on the nose gear of the subject of Profile 50, P-38J Irish, at Nadzab.

CHAPTER 12
36th Fighter Squadron "Flying Fiends"

The 35th, 36th and 80th Fighter Squadrons comprised the 8th FG and had arrived in Australia in late February 1942. The 36th FS fought at Port Moresby and Milne Bay throughout that year, before transferring to northern Queensland in early 1943 for recuperation. In June 1943, and with no Lightnings available, the unit upgraded to new P-39Q Airacobras which it flew in New Guinea mainly on patrols and escorting transports. In November 1943 the squadron commenced transition to P-47Ds which it operated briefly before training on war-weary P-38Hs handed down from the 80th FS at Townsville in early February 1944.

The 36th FS's first P-38 casualty was First Lieutenant Robert Cordell, who was killed during a training flight at Townsville on 25 February 1944. With conversion complete, Captain Donald Campbell was appointed as the new squadron commander on 16 March 1944 to lead the 36th FS throughout its P-38 era in the SWPA. Its last loss in the theatre was on 19 June 1944 when Lieutenant Carl Younggren was killed on a training mission near Gusap when a flaming engine placed him in a spin. The squadron's only known combat loss of a P-38 in the SWPA occurred on 15 March 1944 when Captain Warren Damson was shot down by JAAF fighters near Wewak. The 36th FS took delivery of a batch of new P-38J-15s in April 1944, before transferring out of the SWPA to Dutch New Guinea two months later. It lost eleven P-38s to various non-combat causes in the SWPA, with Dawson representing the sole P-38 combat loss by the squadron.

The 36th FS's biggest loss in one day occurred during the infamous "Black Sunday" mission of 16 April 1944 when four P-38s were disorientated in capricious weather during an escort flight led by Lieutenant Eugene Zielinski. The task was to escort a solitary 822nd BS B-25G Mitchell on a search of the Sepik plain for the crew of a downed B-25G Mitchell which had put down behind Wewak four days previously. Accompanying Zielinski were Lieutenants Walter Mikucky (pronounced *"Mikusky"*), Robert Keown (pronounced *"Cowan"*) and Lawrence Reeves (flying a borrowed 80th FS Lightning). All four Lightnings were claimed by the weather.

Markings

The 36th FS's Lightnings were numbered 1 through to 29, a departure from its Airacobra days where their fighters were assigned an alphabetical letter. The squadron painted the tailplane tips of its natural metal finish Lightnings white with black piping, and applied a similar scheme to its spinners, all of which varied slightly from fighter to fighter.

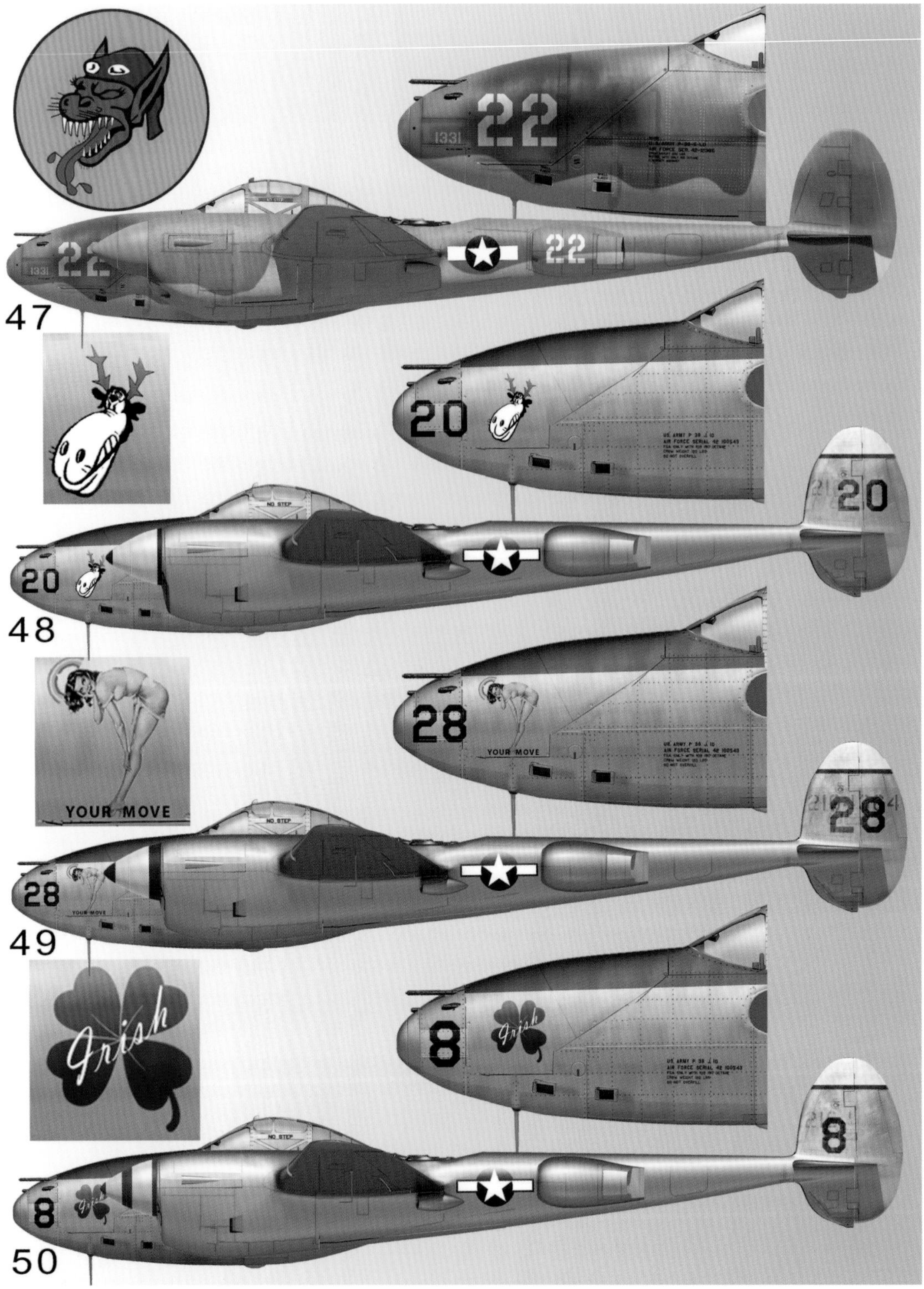
22
1331
47
20
48
28
YOUR MOVE
49
8
Irish
50

36th FS Logo –A gargoyle-style beast wearing a leather helmet with a blooded tongue and bloodshot eyes.

Profile 47 – P-38H serial 42-66668, squadron #22

This fighter was transferred from the 80th FS where it was named *Corky Jr.* The 36th FS removed the nose-art and painted the squadron number 22 on the nose. The Constructor's Number 1331 remained on the nose in yellow stencils. The markings of this fighter are taken from examination of the wreckage in 2018 which remains extant at its 16 April 1944 crash site. Lieutenant Robert Keown was one of four lost during the squadron's biggest loss in one day during the infamous "Black Sunday" mission. Keown returned to the circuit area of Yamai, where he circled off-shore in poor visibility and rain, then climbed towards the Finesterre Range and disappeared from view. He failed to make Nadzab and instead baled out in heavy weather shortly afterwards. In the 1980s a villager located his remains in a high swamp and buried them in the tribal cemetery. In 2005 the wreckage of his P-38H was located separately.

Profile 48 – P-38J serial unknown, squadron #20, (moose illustration)

This fighter was a new P-38J-15 in the serial 42-1043XX range assigned into the 36th FS in late March 1944. Although it carried no name, its nose was decorated with this moose artwork reflecting squadron commander Captain Donald Campbell's nickname. The serial number on the fins was painted over, but parts became partially visible through weathering. Profiles in other publications unfortunately misrepresent the art work. The outline art here was replicated precisely from a close-up photograph of it. Campbell flew this fighter when he led a dozen of the squadron's Lightnings to escort bombers on "Black Sunday".

Profile 49 – P-38J serial unknown, squadron #28, *Your Move*

This was another new P-38J-15 in the serial 42-1043XX range assigned into the 36th FS in late March 1944. Its assigned pilot is unidentified.

Profile 50 – P-38J serial unknown, squadron #8, *Irish*

This fighter was assigned to Lieutenant Orville Erdmann at Nadzab sometime around May 1944.

Ferocious Ferdinand was an unidentified P-38J-15 assigned into the 36th FS at Nadzab. It replaced P-38H serial 42-66668 (the subject of Profile 47) which was lost on "Black Sunday".

The subject of Profile 52, P-38H serial 42-66893, after it crash-landed at Barakoma in October 1943.

Captain George Seeberg and Lieutenant Lowell Schurr pose with Ski Heil, a play on the phrase "Sieg Heil", at Guadalcanal's Fighter Two in July 1944. Unusually the fighter has white spinners. It is the subject of Profile 56.

CHAPTER 13
44ᵗʰ Fighter Squadron "Vampires"

Activated at Wheeler Field as the 44ᵗʰ Pursuit Squadron on 22 November 1940, the squadron was assigned to the 18ᵗʰ PG and redesignated as the 44ᵗʰ FS on 15 May 1942. The squadron deployed to the New Hebrides in November 1942 and from the following month it operated later model P-40s from Guadalcanal. The two squadron commanders who oversaw the unit's introduction to the Lightning were Major Robert Westbrook, from 25 September 1943, and then Major Peyton Mathis from 6 February 1944. Mathis came to the 44ᵗʰ FS as a veteran of the North African and Italian campaigns and enjoyed the reputation of being the first pilot to fly a P-38 non-stop from England to Africa. He was killed in the South Pacific on 5 June 1944 in a flying accident (see below).

Although the USAAF did not classify the 44ᵗʰ FS as a twin-engine one until 20 February 1944, several of its pilots were seconded to other Lightning units commencing in September 1943, followed by another group who undertook conversion training in New Caledonia two months later. In January 1944 the squadron was relocated to Guadalcanal in order to receive a batch of P-38Hs and early model Olive Drab P-38Js. A detachment of these was soon operating from Munda, and in January 1944 some missions involved up to twenty P-38s escorting bombers to Rabaul.

At the beginning of March 1944, the 44ᵗʰ FS moved up to Stirling Island in the northern Solomons, where the pilots were attached to the 339ᵗʰ FS for advanced training. For the entire month of April 1944, with Japanese fighters having left the theatre, the squadron focused on ground strikes against airfields near Rabaul and Kavieng. With the air war in the South Pacific having scaled down by May 1944, the unit packed up to move to Morotai in the Netherlands East Indies. By January 1945 it was operating from the Philippines.

During its time in the South Pacific, the 44ᵗʰ FS lost eight Lightnings which went missing during operations and two to other causes. At least one pilot, Lieutenant John Cox, became a POW and was later assassinated. The first P-38 combat loss occurred on Christmas day 1943, while the last loss was squadron commander Major Peyton Mathis, lost to an accident at Guadalcanal on 5 June 1944 (see Profile 55).

Markings

During initial P-38 operations the 44ᵗʰ FS operated fighters from other units, mainly the 339ᵗʰ FS. These early Lightnings were allocated squadron numbers commencing at 30, however the system changed around March 1944 when the first natural metal finish P-38J-15s were delivered at Guadalcanal. The numbering system was revised commencing at 400, with 400 reserved for the squadron commander. The numbers were stenciled in yellow with black piping and appeared on the nose gun panel and on the fins. Sometimes the number was also painted on the boom air scoops. Olive Drab Lightnings which remained in the inventory had 400-series numbers painted in white. Lightnings assigned to flight leaders had one stripe, the operations officer had two whilst the commanding officer had three.

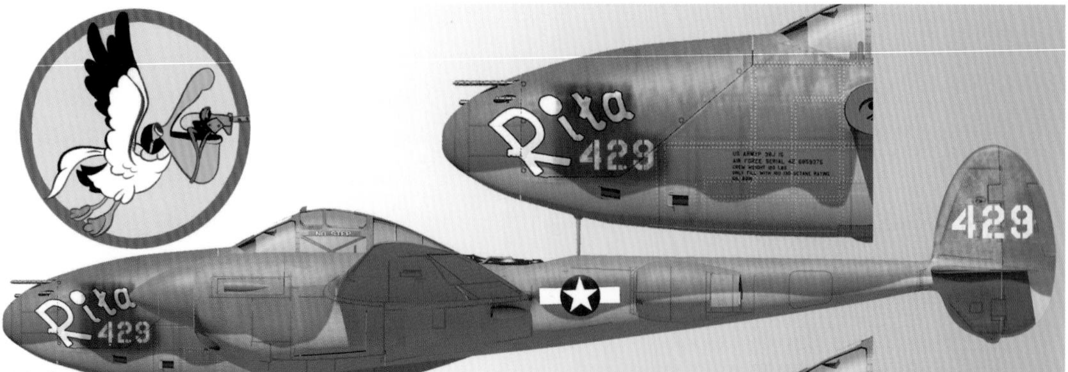
Rita
429
429

51

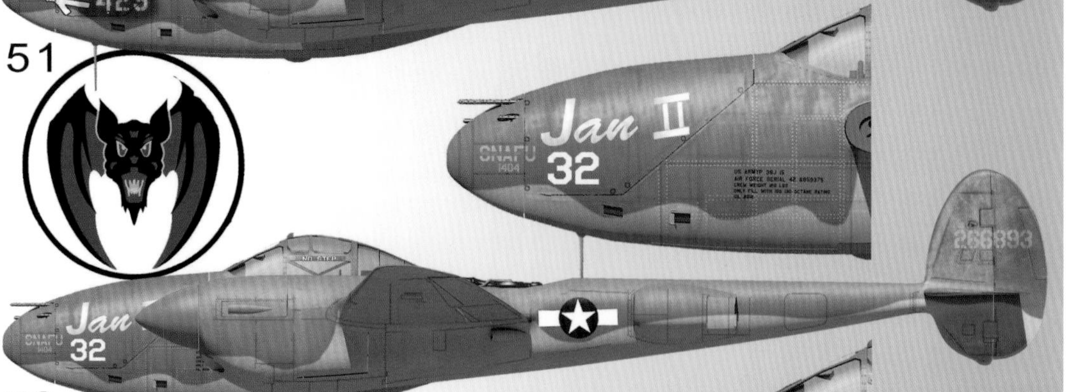
Jan II
SNAFU
1404
32
266893

52

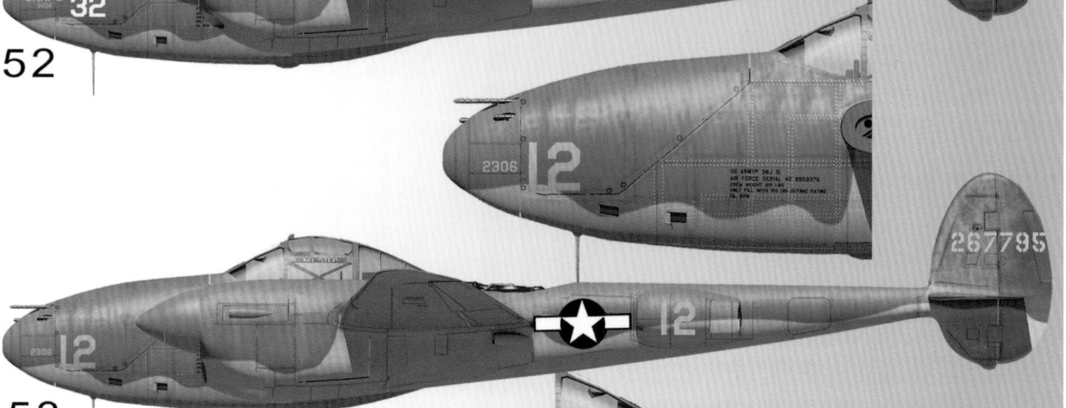
2306
12
267795

53

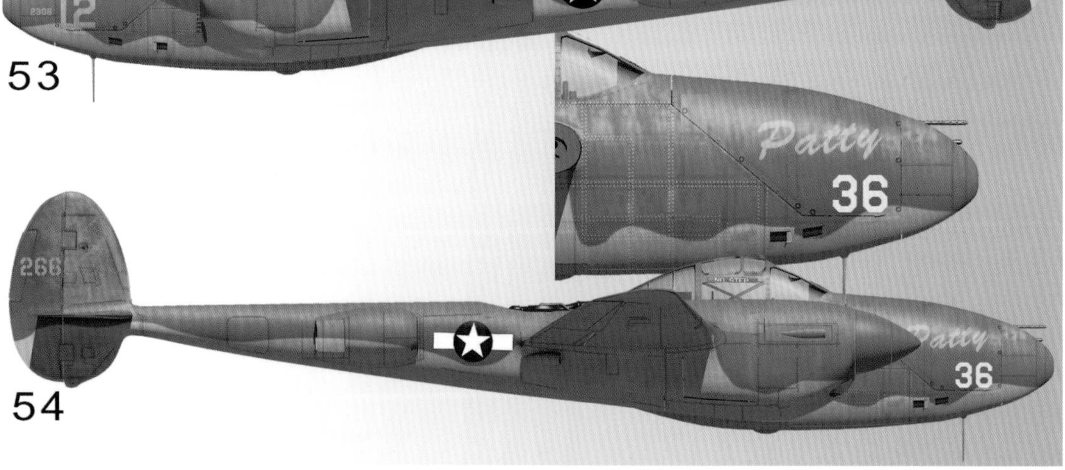
Patty
36
266

54

44th FS Logo

The original 44th FS logo was designed by Walt Disney and featured a critter hiding in a stork's beak, firing a machine gun. Although this logo appears nowhere in the official histories, it was used during the early Guadalcanal campaign and appeared outside unit offices. It was replaced by a second logo of a Vampire face bearing blooded teeth somewhere around mid-1944, and this second logo is most closely associated with the wartime 44th FS.

Profile 51 – P-38G serial unknown, squadron #429, *Rita*

This P-38G was transferred to the 44th FS from the 339th FS, and the serial number was possibly 43-2210.

Profile 52 – P-38H serial 42-66893, squadron #32, *Jan II*

This P-38H was transferred from the 339th FS which had named the fighter *SNAFU*. The Constructor's Number 1404 remained on the nose under the faded original name even after it was renamed in 44th FS service. On 7 October 1943 *Jan II* was force-landed at Barakoma by Lieutenant David Livsey following combat.

Profile 53 – P-38J serial 42-67795, squadron #12

This fighter is illustrated as it appeared at Guadalcanal shortly after its assignment to the 44th FS from an unknown unit. It ended its career serving with the 301st Airdrome Squadron at Nadzab where it was lost to an accident on 13 February 1945.

Profile 54 – P-38H (serial unknown), squadron #36, *Patty*

It is likely this P-38H was received from the 339th FS, and then allocated squadron number 36. This system was revised around March 1944 when the first natural metal finish P-38J-15s were delivered to the 44th FS, replaced by a numbering system which commenced at 400.

Hawkeye Hattie (not profiled) as a brand-new fighter at Guadalcanal in August 1944. Note it has one stripe on the boom denoting a flight leader.

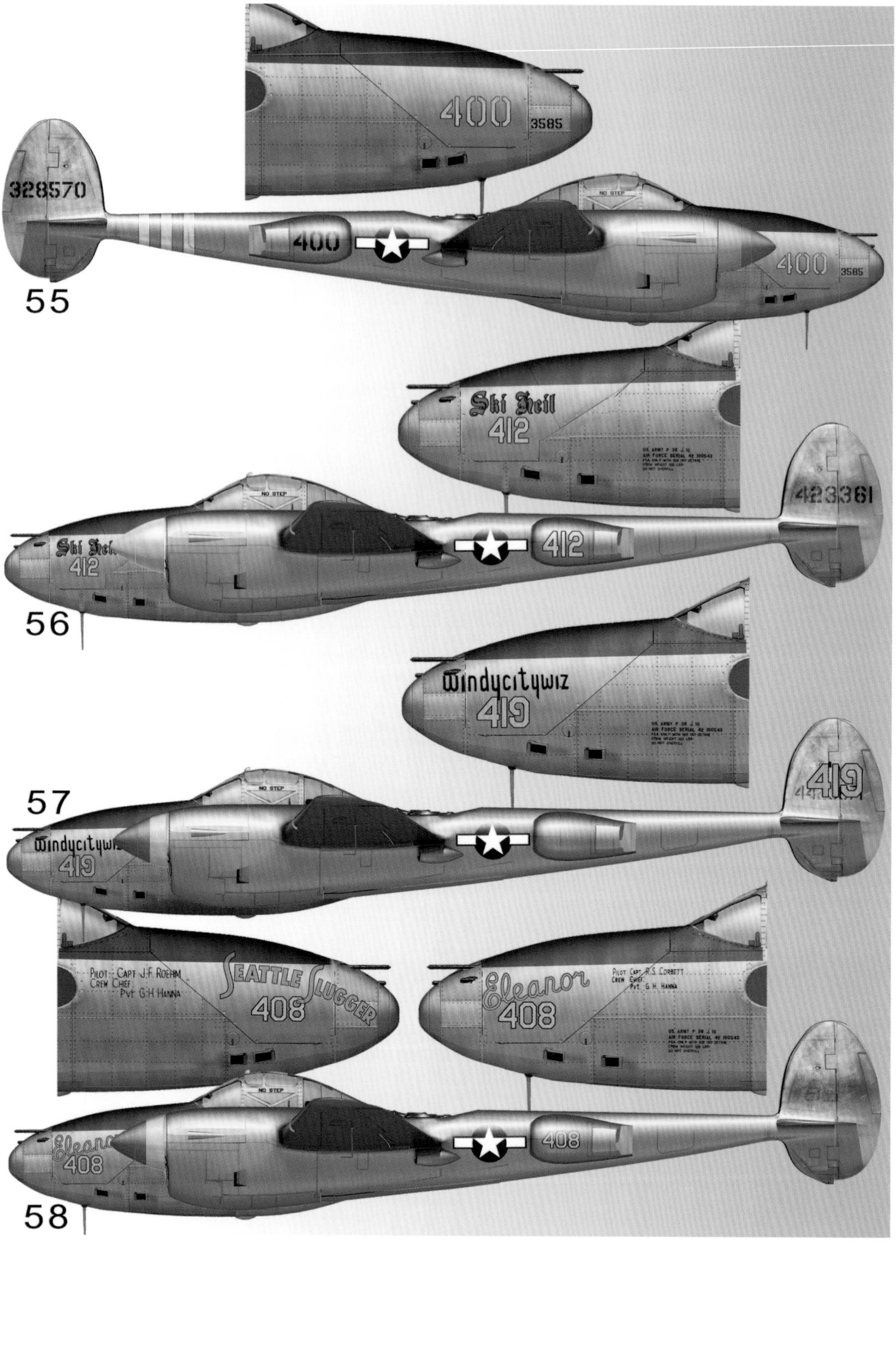
400
3585
328570
NO STEP
55
Ski Heil
412
423361
56
Windycitywiz
419
57
PILOT: CAPT J.F. ROEHM
CREW CHIEF:
PVT G.H HANNA
SEATTLE SLUGGER
408
Eleanor
PILOT CAPT R.S. CORBETT
CREW CHIEF:
PVT. G.H. HANNA
58

Profile 55 – P-38J serial 43-28570, squadron #400

On 5 June 1944 the 44th FS commander Major Peyton Mathis led a dive-bombing mission against Poporang Island; however, the mission was cancelled due to weather. Experiencing engine trouble, Mathis jettisoned his bombs before returning to Guadalcanal on one engine. He remained in formation until he reached Savo Island where he instructed the others to land first. Mathis then made a long final for Fighter Two but at 1035 lost asymmetric control and spun into a ravine six miles east of the airfield. Note the three stripes on the boom denoting his status as squadron commander.

Profile 56 – P-38J serial 44-23361, squadron #412, *Ski Heil*

Derived as a witticism from the phrase "Sieg Heil", this P-38J was assigned to Captain George Seeberg whose surname has an Austrian derivation. It is profiled as it appeared at Guadalcanal's Fighter Two in July 1944, with its unique white spinners.

Profile 57 – P-38J-20 serial unknown, squadron #419, *Windycitywiz*

This fighter was assigned to Lieutenant William Starke who became one of the 44th FS's longest-living pilots when he died at the age of 99 on 6 June 2018.

Profile 58 – P-38J serial unknown, squadron #408, *Seattle Slugger/Eleanor*

This fighter was among the first batch of J models assigned into the 44th FS at Guadalcanal. Note that that the squadron number has been applied on the boom and not the tail.

The subject of Profile 57, P-38J Windycitywiz, at Guadalcanal.

The starboard nose art of Lieutenant Allen Hill's P-38 serial number 42-67776 named Hill's Angels. Past publications have consistently misidentified this fighter whose Constructor's Number 2287 can just be seen, painted over with the squadron letter H.

CHAPTER 14
80th Fighter Squadron "Headhunters"

After disembarking at Brisbane on 6 March 1942, the 80th FS operated Airacobras for its first year in the SWPA alongside the two other squadrons of the 8th FG, the 35th and 36th FS. The squadron later achieved notoriety as the "Headhunter" Lightning squadron in New Guinea, receiving substantial publicity at the time. It began converting to the P-38 in Queensland in mid-February 1943. On 21 March 1943 the 80th FS moved to Three Mile 'drome (Kila) at Port Moresby. Captain Edward Cragg, recently appointed as the squadron commander, took the Lightning equipped unit to New Guinea.

The transition to the much-coveted new type was made under controversial conditions. Originally the 35th FS was slated as the first 8th FG squadron to receive Lightnings; however, the decision was reversed by Brigadier General Ennis Whitehead on the basis that the 35th FS was then hampered by a disproportionate number of malaria cases. Thus, he chose to allocate the Lightnings to the 80th FS as a more combat-effective unit, while the 35th FS would use extra time in Australia to recuperate. Animus between the two squadrons subsequently lingered for a long time over this decision. A bottom line, however, is that there were never sufficient P-38s for the fighter squadrons that needed them.

The 80th FS's first P-38 mission unfolded on 30 March 1943 when four Lightnings escorted RAAF Beaufighters on a strike. Then, during the 12 April 1943 Operation *I-Go* raid against Port Moresby the trio of Lieutenants Donald McGee, Campbell Wilson and Bob Siebentahl launched to meet incoming Japanese bomber formations. Wilson's starboard engine faltered on the outbound leg, so he headed back to Port Moresby but near Yule Island a solitary Zero shattered his canopy. Wilson then made a wheels-up landing at 30-Mile airfield.

On 4 June 1943 Lieutenant George Welch (of Pearl Harbor fame) was appointed as the 80th FS Operations Officer. The squadron first operated P-38Gs, gradually bolstered by previous 39th FS P-38Fs and several P-38Hs. A batch of new Olive Drab P-38J-5s was assigned to the unit at the end of 1943, followed by the first natural metal finish P-38Js around the end of March 1944.

The 80th FS was involved in some of the heaviest fighting in the theatre and was a regular participant in missions against both Wewak and Rabaul. On 21 August 1943 the squadron lost four Lightnings on a combat mission escorting B-25s to Wewak, and then three over Rabaul during the infamous 2 November 1943 Fifth Air Force combined attack. The last loss in the SWPA theatre occurred on 7 May 1944 when the nose wheel of *Jandina III* failed to extend and the fighter was written off while landing at Yamai on New Guinea's northern coast. In June 1944 the unit departed for Dutch New Guinea.

During its time in the SWPA the 80th FS lost a total of 41 Lightnings: 17 in combat and 24 to various other causes including weather-related losses.

Markings

All 80th FS Lightnings were allocated an alphabetical identifier with the rare exceptions of at least one issued with a numeral (see Profile 60). Spinners and tail tips were painted with the squadron colour green.

P-38J 42-67587 Princess Pat II seen at Nadzab. This aircraft replaced Princess Pat which is depicted in Profile 61. Note that the letter G has been painted over the tail serial, rather than removing it first.

P-38J-15 Corky IV at Nadzab on 4 May 1944. This was one of the first natural metal finish models assigned to the 80th FS and was allocated to Lieutenant Cornelius "Corky" Smith.

Lieutenant Kenneth Ladd with the subject of Profile 60, P-38G serial 43-2197, at Nadzab.

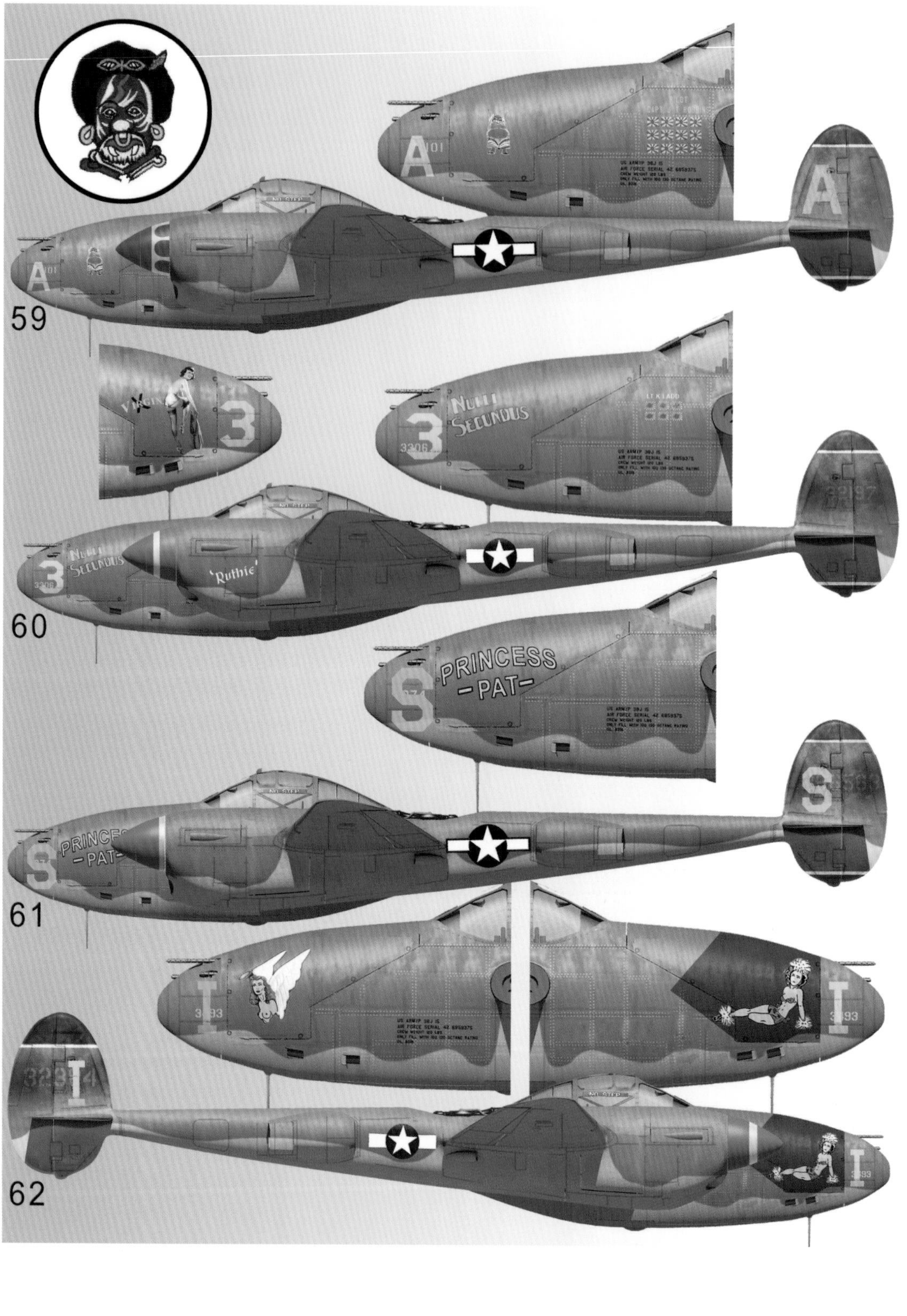
59
A
NULLI SECUNDUS
3306
3
'Ruthie'
60
PRINCESS
-PAT-
S
61
I
62

80th FS Logo – A New Guinea "headhunter" cannibal adorned with facial warpaint.

Profile 59 – P-38J serial 42-67590, squadron letter A, *Jandina II*

This Olive Drab J model was assigned to Lieutenant Jay Robbins and crew chief Staff Sergeant Devereaux. Named by Robbins as his second assigned Lightning, the nickname derives from a spliced combination of his name "Jay", "and" and his wife's nickname "Ina". Although it was known as *Jandina II*, the name did not appear on the airframe. On the port side of the gondola appears a small, robed Buddha with both hands raised. The fighter is illustrated with a scoreboard of a dozen rising sun flags. Robbins flew this fighter until he traded it in for an natural metal finish P-38J-15 in late March 1944, which he named *Jandina III*.

Profile 60 – P-38G serial 43-2197, squadron #3, *XVirgin/Nulli Secundus*

This Lightning was assigned to Lieutenant Kenneth Ladd who named it *Nulli Secundus* which translates in Latin as "second to none". The fighter's crew chief was Sergeant Yale Saffro. It suffered a bad accident on 24 August 1943 and was sent to a service squadron for repair before reassignment to the 9th FS. It is unclear why this fighter had a numeral applied as an identifier instead of an alphabetical letter, applied over the Constructor's Number 3306 on the nose.

Profile 61 – P-38H serial 42-66563, squadron letter S, *Princess Pat*

Assigned to Lieutenant Don Hanover and crew chief Sergeant Joe Clark, this fighter went missing on 2 September 1943 during combat with JAAF fighters over Wewak. Flown by Lieutenant Robert Adams, it was escorting 38th BG B-25 Mitchells on a strafing mission against shipping off Wewak. The Lightning was replaced by P-38J 42-67587 which was named *Princess Pat II*, but which was allocated the different squadron identifier G.

Profile 62 – P-38H serial 43-2384, squadron letter I

This P-38G was among the first batch of Lightnings assigned into the 80th FS, and although it carried no name it was decorated with different art on both sides of the gondola. It was written off by Lieutenant James Wilson during a take-off accident at Kila 'drome on 15 October 1943.

The proud new owners of P-38J serial 42-104013, the subject of Profile 65, in early April 1944 shortly after its receipt at Nadzab.

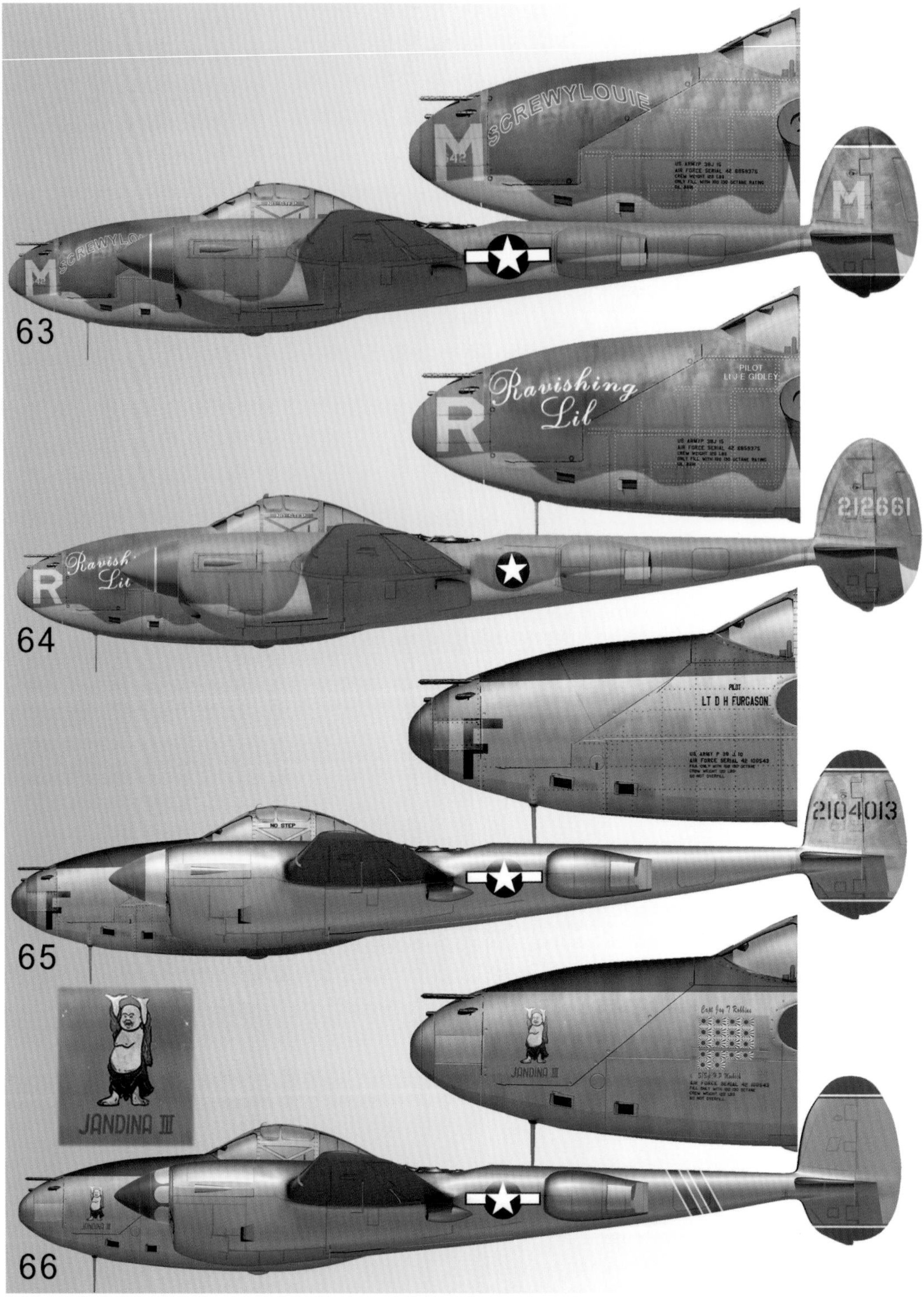
SCREWYLOUIE
M
63
Ravishing Lil
PILOT
R
212661
64
PILOT
LT D H FURGASON
NO STEP
2104013
65
JANDINA III
66

Profile 63 – P-38J serial 42-67131, squadron letter M, *Screwy Louie*

This fighter was assigned to and named by Lieutenant Louis Schriber. It survived its combat tour and was eventually scrapped in New Guinea.

Profile 64 – P-38F serial 42-12661, squadron letter R, *Ravishing Lil*

This fighter was received by the 80th FS on 17 September 1943 and later assigned to Lieutenant Jess Gidley. Major Ed Cragg claimed a Zero on 24 October 1943 with this fighter which survived its combat tour and was struck off charge on 21 December 1944.

Profile 65 – P-38J serial 42-104013, squadron letter F

This natural metal finish P-38J was assigned to Lieutenant Delbert Ferguson in late March 1944 at Nadzab. He secured the squadron letter F to represent his surname.

Profile 66 – P-38J serial 42-103988, *Jandina III*

This natural metal finish J model succeeded *Jandina II* (see Profile 59) when it was assigned to Captain Jay Robbins (recently promoted) and his crew chief Staff Sergeant HP Mosback at Nadzab. On 7 May 1944 Robbins headed to Saidor to refuel following a patrol of New Guinea's northern coast. After experiencing a loss of nose wheel hydraulic pressure which prevented him from lowering the nose gear, he force landed at Yamai airfield near Saidor. The P-38J was written off and placed on drums where it was stripped for parts. The airframe was salvaged from the site in 2002. The scoreboard shows 19 kills, and the names of both Robbins and Mosback were painted in flowing red script. The fighter had no alphabetical identifier applied. The intricate decoration of the red, white and blue spinners reflect the colours of the US flag.

Major Edward Cragg's Porky II at Kila 'drome. This P-38H was lost on 26 December 1943 over New Britain. Although Cragg was declared Missing in Action, the diary of gunner Kubota Yoshio serving with an Independent Field Artillery unit at Cape Gloucester reveals Cragg's capture on 2 January 1944. There is no record of his subsequent fate, however it is clear he did not survive to be transported to Rabaul.

Following the "Yamamoto Mission", P-38 squadron number 102 (serial 42-12708) was seen by a PBY crew south of Shortland Island at 0940 trailing vapor from a feathered port engine flown by Lieutenant Howard Hines. The PBY radioed Hines, who reported that he was OK and had sufficient fuel to return to base. He asked for the heading for Guadalcanal, and then banked away but was never seen again. Hines remains Missing in Action, and this is the last known photo of 102 on Guadalcanal.

The subject of Profile 69, P-38G serial 43-2264 Miss Virginia, at Henderson Field with the Constructor's Number 3373 just visible on the nose.

CHAPTER 15
339th Fighter Squadron "Sunsetters"

The 339th FS is famous for leading the "Yamamoto Mission'" on 18 April 1943 with P-38G Lightnings. It had commenced combat in Airacobras and also undertook the first night fighter missions in the South Pacific. Major Dale Brannon was appointed commanding officer when the squadron was established in New Caledonia on 29 September 1942, on the understanding that the unit would be the first to operate Lightnings in the theatre.

An advance detachment of seven pilots was sent to Guadalcanal in September 1942 where they shared billets with the 67th FS and briefly flew Airacobras. They returned to New Caledonia to train on the first P-38Gs to arrive in the theatre which were unloaded in New Caledonia in October. The P-38 first appeared on Guadalcanal on 12 November 1942, eight of them ferried over led by Brannon. They were briefly joined by an attachment of eight on temporary assignment from the SWPA the following day operated by the 39th FS (see Chapter 3).

In November 1942 Captain John Mitchell replaced Brannon as commander when the 339th FS commenced P-38 operations on Guadalcanal. Its unofficial nickname soon became the "Sunsetters" due their dabbling in the use of P-38s as night fighters in conjunction with Guadalcanal's searchlights (see Chapter 4). The squadron's first loss occurred on 15 December 1942 when Lieutenant Eugene Woods ditched when returning from a Munda strike. Although observed floating in the water in his life jacket, he remains Missing in Action.

On 5 January 1943 Mitchell led six Lightnings (along with 68th FS Warhawks) in escorting 11th BG B-17Es over Tonolei Harbor in southern Bougainville. They were met by Zeros alongside Pete and Rufe floatplanes. From this fight the Lightnings claimed three Japanese aircraft, however only one Rufe was downed and the 339th FS lost two Lightnings.

On 23 February 1943 Pentagon paperwork caught up with the squadron which was redesignated as the 339th FS (Twin Engine) although it had already been operating Lightnings for several weeks. On 18 April 1943 Mitchell led the "Yamamoto Mission", which downed two Betty bombers and from which one Lightning remains missing. The squadron later advanced up the Solomons chain including to Munda and Barakoma.

On 29 December 1943 the 339th FS returned to New Caledonia for rest and refurbishment. It returned to the theatre on 15 January 1944, basing itself at Stirling Island from where it flew its last missions in July 1944. After this it moved to Dutch New Guinea and then the Philippines.

The 339th FS operated P-38F, G and H models before taking possession of its first natural metal finish P-38J-15s in late March 1944 at Guadalcanal. Despite its reputation as a crack unit, the squadron lost 46 Lightnings during its time in the South Pacific from November 1942 until July 1944, of which 31 were combat losses. Despite claims of high kill ratios, when these losses are compared to Japanese unit logs, the 339th FS kill ratio against Japanese fighters averages around one to one.

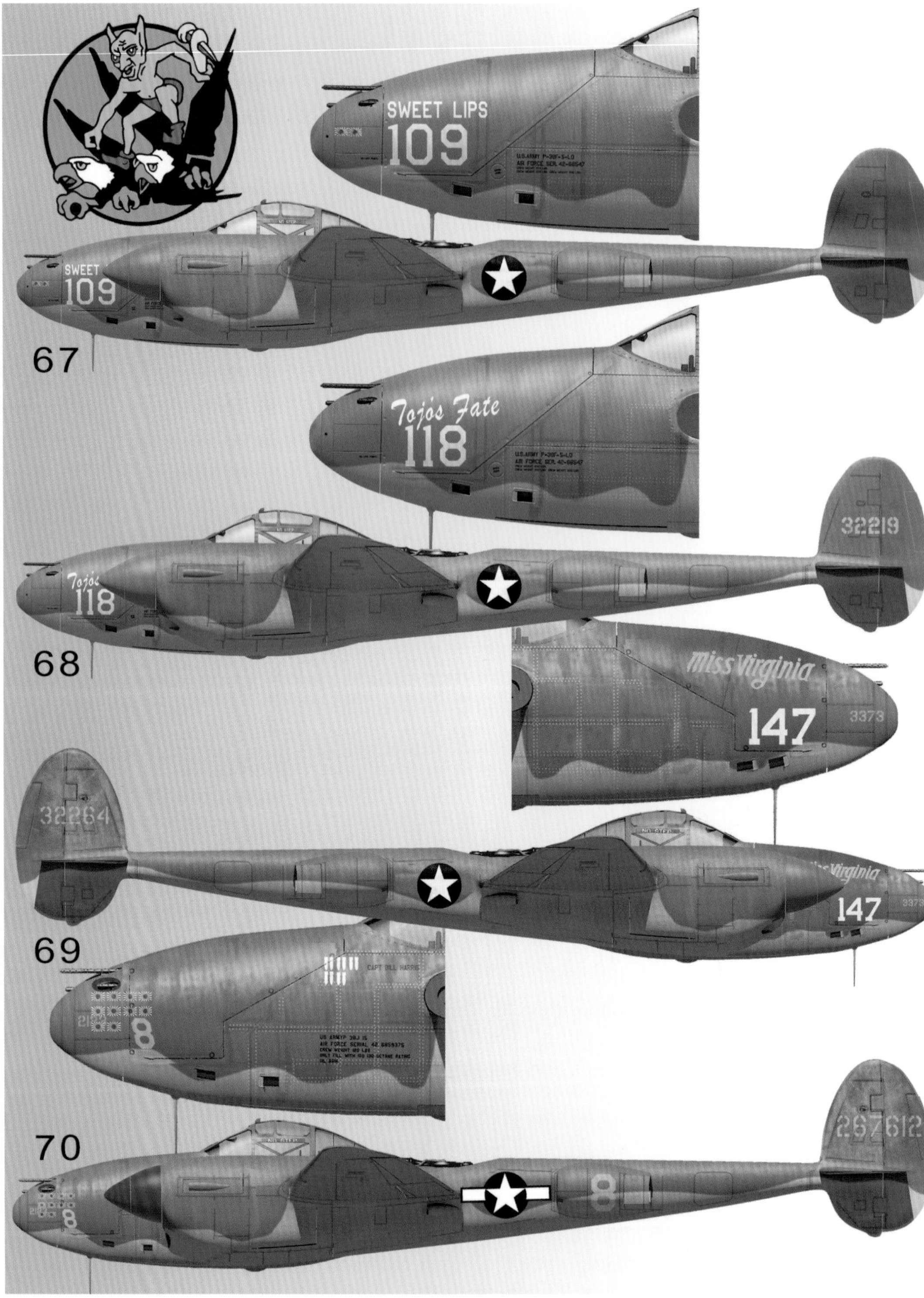
SWEET LIPS
109
67
Tojo's Fate
118
32219
68
Miss Virginia
147
3373
32264
69
CAPT BILL HARRIS
8
70
267612

339ᵗʰ FS Logo – A gremlin riding two eagles to reflect the Lightning's two engines.

Profile 67 – P-38G serial unknown, squadron #109, *Sweet Lips*

This fighter was among the first batch assembled at New Caledonia and was a P-38G-1 in the 42-12XXX range. It was assigned to Lieutenant Earl Conrad who also named it.

Profile 68 – P-38G serial 43-2219, squadron #118, *Tojo's Fate*

This Lightning underwent an accident near Oua Tom on New Caledonia after being assembled. It was repaired and later served on Guadalcanal.

Profile 69 – P-38G serial 43-2264, squadron #147, *Miss Virginia*

This fighter was flown by Lieutenant Rex Barber on the "Yamamoto Mission" over Bougainville on 18 April 1943. The Constructor's Number 3373 appears in yellow stencils on the nose.

Profile 70 – P-38J serial 42-67612, squadron #8

This Lightning was among the first J models assigned into the 339ᵗʰ FS. It had served briefly with the 12ᵗʰ FS before being assigned to Captain Bill Harris, who later became the commander of the entire 18ᵗʰ FG. Note the Constructor's Number 2123 stenciled on the nose.

The subject of Profile 68, P-38G serial 43-2219 Tojo's Fate, being collected after its accident on New Caledonia.

P-38G serial 43-2238, squadron number 122 Phoebe, was flown by Lieutenant Thomas Lanphier on the "Yamamoto Mission".

A 418th NFS field-modified P-38G night-fighter climbs out from Port Moresby at dusk to intercept a radar-plot of incoming G4M1 bombers from either Kavieng or Rabaul.

CHAPTER 16
418ᵗʰ Night Fighter Squadron

Activated on 1 April 1943, this dedicated night fighter unit docked at Brisbane on 29 October 1943 aboard the USS *General John Pope*. It then proceeded to Townsville where it boarded the SS *Van Huestz* which sailed to Dobodura, arriving on 22 November. The unit, commanded by Major Carroll Smith, answered directly to Fifth Fighter Command and operated five P-70s and two P-38s converted to night fighters. Both P-38s were field-modified with radar at Port Moresby. Four more unconverted P-38Js were later assigned along with four F and G models from the 475ᵗʰ FG inventory. On 28 March 1944 the 418ᵗʰ NFS moved to Finschhafen.

With limited scope for night interceptions, the 418ᵗʰ NFS also engaged in daytime combat missions, such as when it lost its first P-38 on 16 January 1944. This occurred when Lieutenant William Carriger caught a wingman's prop-wash during a strafing run, causing his fighter to stall on one wing and crash.

The 418ᵗʰ NFS took up night alerts and patrols from Dobodura, Port Moresby, Kiriwina and Finschhafen during its time in the SWPA. On 2 January 1944 two of its P-38s covered the landing convoy at Saidor for two hours commencing at dawn. Tropical weather claimed more losses than combat, especially during extended patrols during which twin drop-tanks were carried. On 5 March 1944 Lieutenant Edward Craig went Missing in Action in the Admiralty Islands, running out of fuel before he could find an Allied base.

From 1 to 13 April 1944 the 418ᵗʰ NFS used its P-38s to run an extensive series of night fighter patrols between Cape Croiselles and Kar Kar Island. It exited the SWPA when it moved to Hollandia in Dutch New Guinea on 12 May 1944. During its brief four months operating in New Guinea the squadron lost four P-38s to various causes including one to "friendly" anti-aircraft fire.

The subject of Profile 71, P-38G serial 42-12851, at Port Moresby in early 1944. (courtesy Chris Narzisi)

418th NFS Logo

The "Kingbee" squadron logo was a black and golden orange bee, wearing a red crown and holding aloft a lighted lantern. The right foreleg represented radar, while the left foreleg grasped a machine gun. Tiptoeing across a cloud base at night, the bee is peering over the edge with a look of ferocity.

Profile 71 – P-38G serial 42-12851, squadron #51

This converted night fighter was received from the 6th NFS around December 1943 when that squadron's New Guinea detachment returned to the US. It was later destroyed in an accident on 9 September 1944. The squadron number was the last two digits of the serial number.

Profile 72 – P-38G serial 42-12705

This Lightning was received from the 80th FS where it had been named *Veni Vidi Vici*. The artwork was painted out. On 4 March 1944 Lieutenant Edward Craig went missing between Manus Island and Finschhafen during a two-aircraft night patrol. Late radio contact was made with Craig about 2030, however both Lightnings were soon caught in a storm. The other Lightning flown by Lieutenant Robert Forrestal eventually force-landed at Cape Gloucester suffering from fuel exhaustion, but Craig remains Missing in Action.

A front aspect view of the subject of Profile 71 at Port Moresby in early 1944. Note the twin long range fuel tanks and the radar antenna on the nose. (courtesy Chris Narzisi)

CHAPTER 17
419th Night Fighter Squadron

Activated on 1 April 1943, the 419th NFS is mostly associated with the P-61 Black Widow, however it also used Lightnings extensively during its first five months of operations. The demand for night-fighters in the South Pacific theatre was accelerated subsequent to a raid by Japanese bombers against Guadalcanal on the night of 20-21 March 1943. Despite attempted interception by P–70s, a total of twenty Liberators and Fortresses was badly damaged by the attack. A USAAF review concluded that more night fighters were required to defend Guadalcanal, and the newly formed 419th NFS was slated to do the job. The squadron arrived in the theatre on 15 November 1943, where it soon became part of the 18th FG. Equipped with ground control radar, the squadron nonetheless struggled to launch just three night patrols, six scrambles, four intruder missions and four daylight sorties by the end of the year.

The 419th NFS allocated squadron numbers 300 to 309 to their first nine Lightnings. Six of the unit's fifteen pilots had previously served with the RCAF in night fighter operations and had been promised the new P-61 Black Widow upon arrival at Guadalcanal. The much-vaunted new type was a long way off, however, and there were still insufficient P-70s to go around. Thus, the 419th NFS operated the Lightnings longer than envisaged. Several accidents occurred as it took a while for pilots to become acquainted to the P-38's high-performance characteristics. A total of 368 night missions were logged on the type about which the squadron historian says:

> … not once did our P-38s come close to an enemy intruder. Our gunners and radar observers were unemployed - "peeved" is a common word used around here.

Low morale was not helped by constant ribbing from nearby Lightning combat units. When the limitations of the P-38 as a night interceptor became apparent, it was instead decided to use the type, along with more recently arrived P-70s, for daytime harassment missions. These ranged as far as Bougainville and Rabaul, staging through Torokina. In April 1944 the squadron acquired several RA-24Bs, the USAAF reconnaissance version of the SBD dive-bomber. These were employed for instrument training to hone night flying skills, thus usefully occupying the squadron's bored pilots while waiting for the long-promised Black Widows. Given their shortage of overall inventory, the unit also operated a couple of B-25s.

Around March 1944 the 419th NFS traded their B-25s for two additional P-38Hs. They painted them painted all-over matt black, added an extra machine gun to the nose and installed a radar unit from a P-70. The main receiver was installed in a drop tank and the antenna installed on the nose. Finally, the first P-61 arrived on 3 May 1944 and thereafter the Lightning soon disappeared from squadron ranks as the new P-61s took their place.

During several months of operations in the South Pacific the 419th NFS operated a total of 21 P-38G, P-38H and P-38J Lightnings. Two of these went missing and six more were lost to ground fire or other causes.

Markings

The 419th NFS used P-38 squadron numbers 301 through 309. These were stenciled in white on the nose, and the original serial numbers were retained on the fins. Initially the Lightnings were left in Olive Drab, however later additions had the entire undersurface painted matt black, leaving the top surface Olive Drab.

419th NFS Logo

A cricket on a cloud holding a lantern and a gun.

Profile 73 – P-38J serial 42-67168, Squadron #301, Piva 'drome, Bougainville, January 1944

This fighter was later transferred to the CRTC at Nadzab, with which it was lost on 12 May 1945 in an accident.

Profile 74 – P-38J serial 42-67788, squadron #305, Piva 'drome, Bougainville, April 1944

On 18 April 1944 Lieutenant Robert MacDonald flew one of three P-38s in a mid-morning strafing mission against coastal guns around Buka in northern Bougainville. During a strafing pass MacDonald's starboard engine was set ablaze by gunfire. He made a sharp three quarters turn into the dead engine then parachuted into the sea. A VP-44 PBY arrived ten minutes later but could not locate him. He remains one of the squadron's two pilots classified as Missing in Action.

The subject of Profile 76, P-38G serial 43-2275, over the Port Moresby area flown by Lieutenant James Watkins. The 116th US Army Hospital can be seen at the bottom of the photograph.

CHAPTER 18
421st Night Fighter Squadron

The 421st NFS sailed from San Francisco on 4 December 1943 aboard the USAT *Sea Pike*. It docked in Townsville on Christmas Day 1943 and then sailed for Milne Bay where the ground section unloaded on 3 January 1944. Commanded by Major Walter Pharr, the squadron arrived in the SWPA without aircraft as they had been told they would be issued inventory when they arrived in New Guinea. The type intended for the squadron was the P-61 Black Widow however the new design would not become available until mid-1944. At the end of January, and still without aircraft, the unit was ordered to move to Nadzab, and an advanced party departed by C-47 to reconnoitre a camp area.

The 421st NFS's first aircraft were a pair of hand-me-down P-38Fs from the 9th FS delivered on 31 January 1944, followed by a solitary P-70 from the 418th NFS, delivered on 13 February 1944. The P-38Fs were flight-tested from Port Moresby's Kila 'drome. On 16 February two more P-38Gs were delivered, and a fifth three days later. All the P-38s were then flown over the Owen Stanley Ranges to Nadzab. On 21 February fourteen of the unit's pilots, nearly the entire unit aircrew cadre, went to Townsville where they collected five new P-38s to deliver to Nadzab, however disappointingly these were destined for combat units.

On 29 February the 421st NFS flew its first combat mission: three P-38Fs departed Nadzab to stage through Finschhafen in the early morning. Their mission was to cover US landings in the Admiralties however the weather was bad, and one Lightning aborted. Meanwhile Lieutenant Paul Zimmer's gear would not retract fully, and shortly afterwards his port engine governor failed just before the starboard engine lost power. He parachuted some 50 miles up the coast from Finschhafen and was safely back at base that evening. This marked the unit's first P-38 loss in the SWPA.

The 421st NFS commenced flying weather reconnaissance missions on 13 March 1944 along the northern New Guinea coastline. A curious loss occurred on 24 March: for the past several weeks ace Lieutenant Richard Bong had accompanied the unit's fighters on these reconnaissance missions. On this morning's mission Lieutenant Tom Malone borrowed Bong's decorated plane *Marge* (see Profile 93). Tasked with a reconnaissance of Wewak, half an hour into the flight at 30,000 feet Malone experienced engine trouble. Unable to feather the propeller, he baled out well above the overcast and unclear of his position, a full account of which appears in Chapter 22.

On 8 April 1944 Lieutenant Alexander Kuzmick became the 421st NFS's first fatality after a landing accident at Nadzab when flying one of the original P-38Fs. Then, on 19 April, the unit's three remaining P-38Fs were transferred out of the squadron. The 421st NFS continued operations for several more months with P-70s before finally taking delivery of its first P-61s on 1 June 1944. The unit lost three Lightnings in the SWPA, all to non-combat causes.

421st NFS Logo

In June 1943 a competition was held to design a squadron insignia, with the winner being the wife of Lieutenant John Olley, the Chief Engineering Officer. The prize was a dinner for two at the Orange Court Hotel in Orlando, Florida. The original logo depicts Bugs Bunny riding a P-70, and the 1945 version replaced the P-70 with a P-61 Black Widow.

Profile 75 – P-38F serial unknown

This Lightning was transferred to the 421st NFS at Nadzab from the 9TH FS. The spinners were painted in the patriotic colours of the US flag and a unique shark's teeth design was applied around the engine intakes.

Profile 76 – P-38G serial 43-2275, squadron #84

This P-38G was transferred to the 421st NFS from the 80th FS at Port Moresby. Note the overlap of the red spinner onto the engine cowl. The squadron number 84 is taken from the last two digits of Constructor's Number 3384. This fighter was later transferred to CRTC, where its later markings are illustrated in Profile 100.

The subject of Profile 75 at Nadzab.

Major George Prentice shortly after being appointed 475ᵗʰ FG commander in May 1943. The backdrop is his second Double Trouble, P-38F serial 42-12652, then assigned to the 431ˢᵗ FS.

CHAPTER 19
431ˢᵗ Fighter Squadron "Hades Squadron"

The 475ᵗʰ FG, comprising the 431ˢᵗ, 432ⁿᵈ and 433ʳᵈ FS, was activated on 14 May 1943, and adopted the group nickname "Satan's Angels". It became the first Fifth Air Force fighter group which was fully P-38 equipped. With considerable fanfare and subsequent publicity, it soon became the most famous of all the Fifth Air Force Lightning units, and also hosted the most colourful Lightning nose art in the theatre.

The shortage of P-38 deliveries for the Fifth Air Force, exacerbated by a high rate of combat and other losses, saw both the 9ᵗʰ and 39ᵗʰ FS relinquish all their P-38s to the 475ᵗʰ FG at the end of 1943. This rotation of fighters helps explains the complex and interwoven nature of Lightning markings in the latter half of 1943. Created from a select cadre of experienced pilots, drawn controversially from other Fifth Air Force squadrons, the elite unit first assembled at RAAF Amberley in Queensland. This was under the leadership of commander Major George Prentice, promoted from his previous position as commander of the 39ᵗʰ FS (see Chapter 3).

In August 1943 the 475ᵗʰ FG moved to Port Moresby. In mid-March 1944 at Nadzab, the group received its first natural metal finish P-38J-15s whose extra fuel tanks enabled them to reach Hollandia. On 15 May 1944 the group moved to Hollandia, effectively leaving the SWPA.

By 11 August 1943 the 431ˢᵗ FS was operating from Port Moresby's 12-Mile 'drome. Along with the 432ⁿᵈ FS, the squadron was scheduled for their first mission the following day, as an escort for C-47 transports to Tsili Tsili, however they were stood down due to bad weather. The following day Lightnings from both squadrons escorted the transports to Bena Bena, delivering troops to the New Guinea highlands. On 16 August the squadron departed alongside 348ᵗʰ FG P-47Ds escorting C-47s taking supplies to Tsili Tsili, where they were opposed by JAAF 59ᵗʰ *Sentai* Ki-43-IIs. The 431ˢᵗ FS's pilots claimed twelve kills whilst the P-47s claimed three, however only two Japanese fighters were lost. For the rest of the month the squadron mainly escorted bombers to Wewak.

On 22 September the 431ˢᵗ FS covered the Allied landing at Finschhafen with two flights which attacked No. 751 *Ku* Betty bombers. For most of October the squadron escorted bombers on barge hunts along the northern New Guinea coast. On 12 October, the first large-scale attack against Rabaul, the squadron was among 55 of the 475ᵗʰ FG's P-38s which staged through Kiriwina to escort B-25s attacking Rabaul's airfields. On 17 October 431ˢᵗ FS pilot Lieutenant Thomas McGuire had a narrow escape over Oro Bay when shot down by Zeros. He ditched safely but was rescued by the tender USS *Hilo*. On the biggest Fifth Air Force attack against Rabaul of 2 November, the 431ˢᵗ FS lost three P-38s although Lieutenant Owen Giertson was later rescued.

From late November until the end of 1943 the entire 475ᵗʰ FG continued to escort bombers to targets along New Guinea's northern coast including Alexishafen, Wewak and Madang. On 26

December during the Allied landing at Arawe, the 431st FS led by Captain Thomas McGuire, alongside several other USAAF units, intercepted a substantial combined JAAF fighter and bomber formation. Overall USAAF claims were an astounding 60 aircraft against the reality that only one 78th *Sentai* Ki-61 fighter and three Ki-49 bombers were lost to combat. The day marks the most outlandish example of over-claiming by the USAAF in the SWPA theatre.

In early January 1944 the 475th FG's Lightnings flew numerous support missions for the Allied landing at Saidor, whilst continuing attacks against the four major Japanese air bases in the Wewak area. By the end of February 1944, it was rare to find enemy aerial opposition in the skies above New Guinea and Rabaul.

On 11 February 1944 the entire 475th FG escorted B-24s to Kavieng thus completing the longest over-water flight by fighters in the SWPA to date. Several more escort missions to Kavieng proceeded later that month. The 431st and 433rd FS moved to Finschhafen on 26 February 1944, whilst 475th FG headquarters and the 432nd FS remained at Dobodura until 24 March 1944. Shortly afterwards the entire group moved again, this time to Nadzab. On "Black Sunday" of 16 April 1944, the group suffered its greatest loss of the war in bad weather, with the 431st FS losing one Lightning, the 432nd FS two and the 431st FS eight.

On 30 March 1944 the 431st and 432nd FS escorted B-24s to Hollandia for the first time. Two more strikes were made against Hollandia on 1 and 3 April, and for the month's remainder the 475th FG's Lightnings flew nine more escort missions to the Hollandia region, including five strikes against Wadke Island, further westwards from Hollandia along the Dutch New Guinea coast. By 15 May 1944, when the entire group moved to Hollandia, the 431st FS had lost 37 Lightnings in the SWPA: 22 to combat, and 15 to various other causes.

Markings

The 431st FS was allocated numbers 100 to 139. The squadron colour red was applied to spinners and tail tips, with white piping often applied later in the war.

The artwork on P-38H serial 42-66825 as illustrated in Profile 77.

Captain Vince Elliot with the subject of Profile 79, P-38H serial 42-66666, at Dobodura.

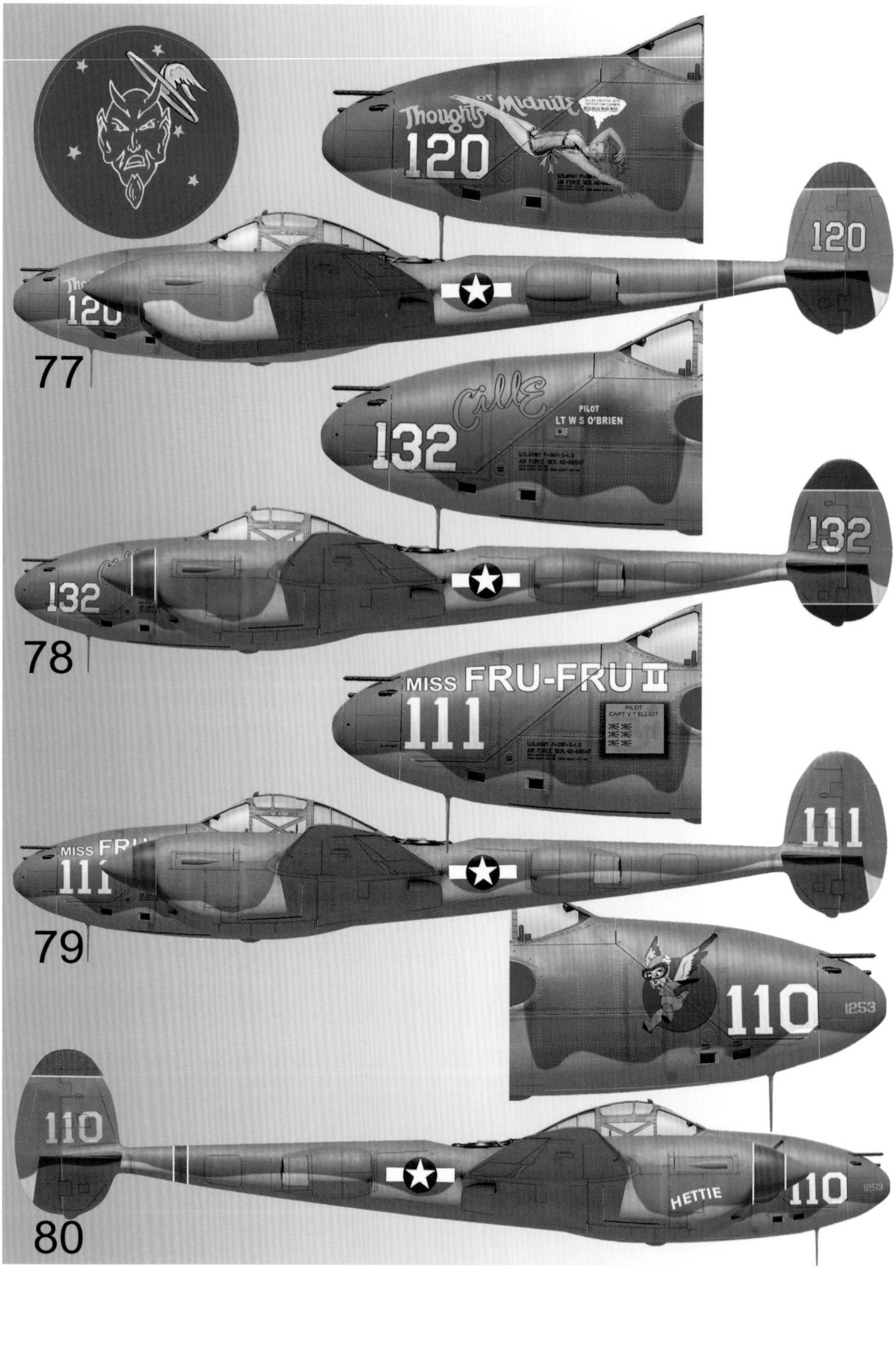
Thoughts of Midnite
120
77
Cille
PILOT
LT W S O'BRIEN
132
78
MISS FRU-FRU II
PILOT
CAPT V T ELLIOT
111
79
110
1253
HETTIE
80

431st FS Logo

A red head of Satan with the stars of the Southern Cross against a blue sky.

Profile 77 – P-38H serial 42-66825, squadron #120, *Thoughts of Midnite*

This fighter was belly landed at Dobodura by its assigned pilot Captain Robert Herman on 2 January 1944 and struck off charge. The fighter's crew chief was Technical Sergeant Charles Corbin.

Profile 78 – P-38H serial unknown, squadron #132, *Cille*

This Lightning was assigned to and named by Lieutenant William O'Brian with crew chief Technical Sergeant Edgar Childs.

Profile 79 – P-38H serial 42-66666, squadron #111, *Miss Fru-Fru II*

The unusual serial number with five 6s was attended by an unusual squadron number of three 1s. The fighter was assigned to and named by Captain Vincent Elliot who brought it in for a crash-landing at Nadzab following combat damage on 16 November 1943.

Profile 80 – P-38H serial 42-66744, squadron #110, (*Fifinella* logo)

The Constructor's Number 1253 appears on the nose in yellow stencils. This fighter was also named *The Woffledigit* on the port gondola and was reassigned to Major Verl Jett when he became 431st FS commander in November 1943. Jett changed the squadron number from 120 to 110, as 110 was reserved for the commander. The art on the starboard side as illustrated originates from the Disney character Fifinella, a female gremlin for a cartoon sourcing from Ronald Dahl's book *The Gremlins*. During WWII, the Women Airforce Service Pilots used the image as their unit mascot. The name *Hettie* on the starboard engine cowl was likely applied by a crew chief.

Profile 81 – P-38H serial 42-66764, squadron #130, *Piss Pot Pete*

This Lightning was assigned to and named by Captain Donno Bellows with crew chief Master Sergeant Gwynne White.

Profile 82 – P-38J serial 42-104032, squadron #134, *T.Rigor Mortis*

This Lightning was among the first batch of natural metal finish J models assigned into the 431st FS in late March 1944 and was assigned to and named by Lieutenant Frank Lent.

Captain Thomas McGuire with P-38J-15 Pudgy III at Nadzab. Five of McGuire's Lightnings eventually bore this nickname for his wife.

The cowl art on P-38H serial 42-66764, the subject of Profile 81.

*The Constructor's Number 2101 on the nose indicates this was previously Lieutenant Jay Robbins'
Jandina II of the 80th FS (see Profile 59). It is seen here shortly after it was transferred to the 432nd FS.
Note the letter A has been painted over, and the unique pattern which attended Jandina II's spinner is
just discernible through the covering white paint on the rear half. A four-leaf clover representing Clover
Squadron has been painted on the nose.*

CHAPTER 20
432ⁿᵈ Fighter Squadron "Clover Squadron"

Note: A brief history of parent group of the 432ⁿᵈ FS,
the 475ᵗʰ FG, is found in Chapter 19.

By 11 August 1943 the 432ⁿᵈ FS was operating from Ward's 'drome, Port Moresby. On 22 September it covered the Allied landing at Finschhafen, alongside other Fifth Air Force fighter units. The squadron was led by Captain Frederick Harris who dived his flight of four into escorting Zeros intending to scatter them so that the other two 431ˢᵗ FS flights could attack the bombers undistracted. Two 432ⁿᵈ FS pilots, Lieutenants Donald Garrison and Vivian Cloud were shot down, although Cloud was later rescued by a destroyer. During the biggest Fifth Air Force attack against Rabaul of 2 November, the squadron lost one P-38.

From late November until the end of 1943 all three 475ᵗʰ FG squadrons escorted bombers to targets along New Guinea's northern coast including Alexishafen, Wewak and Madang. In early January 1944 the entire group flew numerous support missions for the Allied landing at Saidor, whilst continuing attacks against the four major Japanese air bases in the Wewak area. By mid-February 1944 it was rare to find enemy opposition. On 11 February 1944 the group escorted B-24s to Kavieng thus completing the longest over-water flight by fighters in the SWPA to date. The 432ⁿᵈ FS flew more escort missions against Kavieng during the month.

On 25 February 1944 the 475ᵗʰ FG's new station became Finschhafen. On "Black Sunday" of 16 April 1944, the group suffered its greatest loss of the war in bad weather, with the 432ⁿᵈ FS losing two Lightnings. When the 431ˢᵗ FS and the 433ʳᵈ FS moved to Finschhafen on 26 February 1944, the group headquarters and the 432ⁿᵈ FS remained at Dobodura until 24 March 1944. After this date the entire group moved again, this time to Nadzab. On 30 March 1944 the 431ˢᵗ FS and the 432ⁿᵈ FS escorted B-24s Groups to Hollandia for the first time. Two more strikes were made against Hollandia on 1 and 3 April, and for the month's remainder the group's Lightnings flew nine more escort missions to the Hollandia region, including five strikes against Wadke Island, off the New Guinea coast.

The 432ⁿᵈ FS lost 32 Lightnings in the SWPA between August 1943 and May 1944: 14 to combat, two to weather and 16 to other causes.

Markings

The 432ⁿᵈ FS was allocated numbers 140 to 169. The squadron colour yellow was applied to spinners and tail tips, with white piping often applied later in the war.

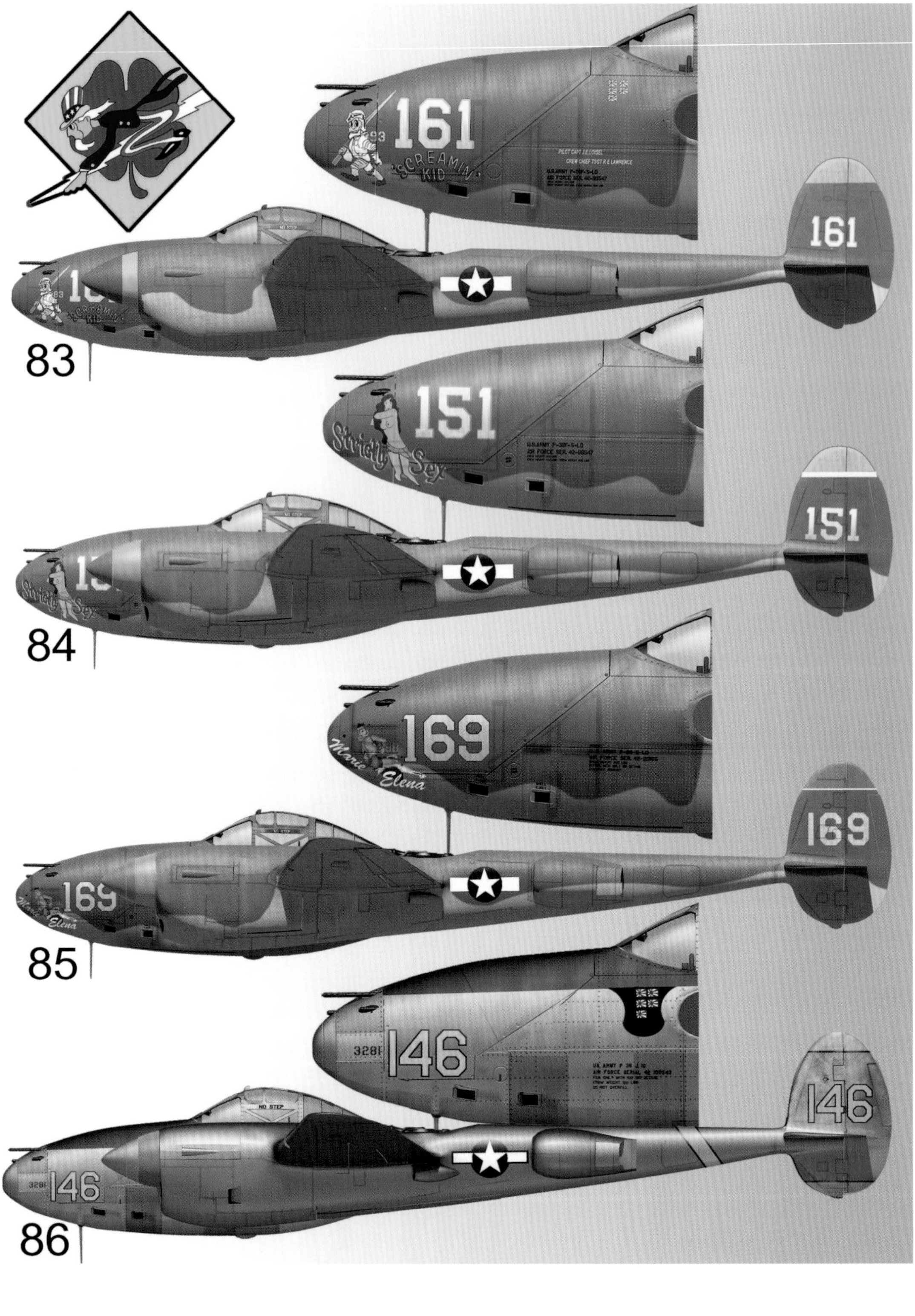
161
SCREAMIN' KID
161
83
151
Strictly Sex
151
84
169
Marie Elena
169
85
146
328F
NO STEP
146
86

432nd FS Logo

Uncle Sam riding a Lightning bolt, in front of a four-leaf clover (the 432nd FS was known as the "Clover Squadron"). The logo was designed by squadron pilot Lieutenant Henry Toll.

Profile 83 – P-38H serial 42-66682, squadron #161, *The Screamin' Kid*

The Constructor's Number 1193 on the nose confirms the correct serial number, incorrectly cited in other publications. This P-38H was assigned to and named by Captain John Loisel, who claimed two Zeros when flying this fighter on 15 October 1943 which was later scrapped in New Guinea in 1945.

Profile 84 – P-38J serial 42-67140, squadron #151, *Strictly Sex*

This Lightning was delivered to New Guinea from Melbourne, arriving on 7 December 1943. It was assigned to and named by Lieutenant Noel Lundy and serviced by crew chief Staff Sergeant Joseph Vogt.

Profile 85 – P-38J serial 42-67787, squadron #169, *Marie Elena*

This Lightning was assigned to Lieutenant Art McLean. Following service in the SWPA, this fighter was lost on 28 July 1944 west of Biak when it ran low on fuel and was ditched by Lieutenant William Elliot.

Profile 86 – P-38J serial 42-104454, squadron #146

This natural metal finish P-38J was assigned into the 432nd FS at Nadzab in late March 1944 and was assigned to Lieutenant Zach Dean. Note the Constructor's Number 3281 on the nose.

P-38J Black Market Babe seen in a Nadzab revetment.

Profile 87 – P-38H serial 42-66734, squadron #161

The subject of Profile 83, P-38H serial 42-66682, with the same squadron number replaced this Lightning after it was lost on 13 September 1943. On that day Lieutenant Noel Lundy departed Dobodura to intercept a JAAF formation heading for Tsili Tsili comprising twelve Ki-49 Helen bombers escorted by 42 Ki-43 Oscars and Ki-61 Tonys. The planned interception was nearly 200 miles distant, and numerous other Fifth Air Force fighter units headed for the same area. Combat became widespread with USAAF claims totaling a dozen victories, whereas Japanese losses were confined to one Tony and two Oscars with no Helens lost. After the combat Lundy headed for home with his wingman when three Tonys ambushed them over the mountains near Bena Bena. Two of the Tonys raked the belly of Lundy's P-38 with accurate fire, and it crashed into mountainous terrain where it still lies today. The markings are referenced from recent photos from the crash site.

Profile 88 – P-38H serial unknown, squadron #160, *El Toro*

This Lightning was named by Lieutenant Ferdinand Hanson after his first name invoking the Walt Disney cartoon character Ferdinand the bull. Some sources cite the serial incorrectly as 42-67290, a P-38J serial number. However, the fighter has the slant cowl of either an F or H model.

The subject of Profile 83, P-38H serial 42-66682 The Screamin' Kid, at Port Moresby shortly after receiving its new nose art, with the last three digits of the Constructor's Number 1193 still evident on the nose.

Lieutenant Ferdinand Hanson (left) with the subject of Profile 88, El Toro, at Nadzab.

Brand new 433ʳᵈ FS P-38H serial 42-66561, squadron number 181, about to be delivered from Amberley to New Guinea. This fighter later went missing later near Buna on 17 October 1943 when flown by Lieutenant Virgil Hagan.

Squadron number 172 undergoing a repair to the leading edge of its port wing at Nadzab.

CHAPTER 21
433ʳᵈ Fighter Squadron "Possum Squadron"

Note: A brief history of parent group of the 432nd FS,
the 475th FG, is found in Chapter 19.

The 433ʳᵈ FS took up station at Port Moresby's Seven-Mile 'drome on 16 August 1943. For the rest of the month, it joined the 475ᵗʰ FG's other two squadrons in escorting bombers to Wewak. During the biggest Fifth Air Force attack against Rabaul of 2 November, the squadron lost only one P-38.

From late November until the end of 1943 the 475ᵗʰ FG escorted bombers to targets along New Guinea's northern coast including Alexishafen, Wewak and Madang. On its first mission to Alexishafen on 9 November 1943, the 433ʳᵈ FS lost Captain Daniel Roberts who had recently taken command of the squadron on 3 October. In early January 1944 the 475ᵗʰ FG flew numerous support missions for the Allied landing at Saidor, whilst continuing attacks against the four major Japanese air bases in the Wewak area. By mid-February 1944 it was rare to find enemy aerial opposition in New Guinea skies. On 11 February 1944 the entire 475ᵗʰ FG escorted B-24s to Kavieng thus completing the longest over-water flight by fighters in the SWPA to date. The 433ʳᵈ FS conducted numerous more escort missions against Kavieng for the month's remainder.

On 25 February 1944 the 475ᵗʰ FG's new station became Finschhafen. The 431ˢᵗ FS and 433ʳᵈ FS both moved to Finschhafen on 26 February 1944, whilst group headquarters and the 432ⁿᵈ FS remained at Dobodura until 24 March 1944. After this date the entire group moved again, this time to Nadzab. On "Black Sunday" of 16 April 1944, the group suffered its greatest loss of the war in bad weather, and the 433ʳᵈ FS lost eight Lightnings.

In total the 433ʳᵈ FS lost 27 Lightnings during its time in the SWPA: eleven to combat, eight to weather and eight to other causes.

Markings

The 433ʳᵈ FS was allocated numbers 170 to 199. The squadron colour blue was applied to spinners and tail tips, with white piping often applied later in the war.

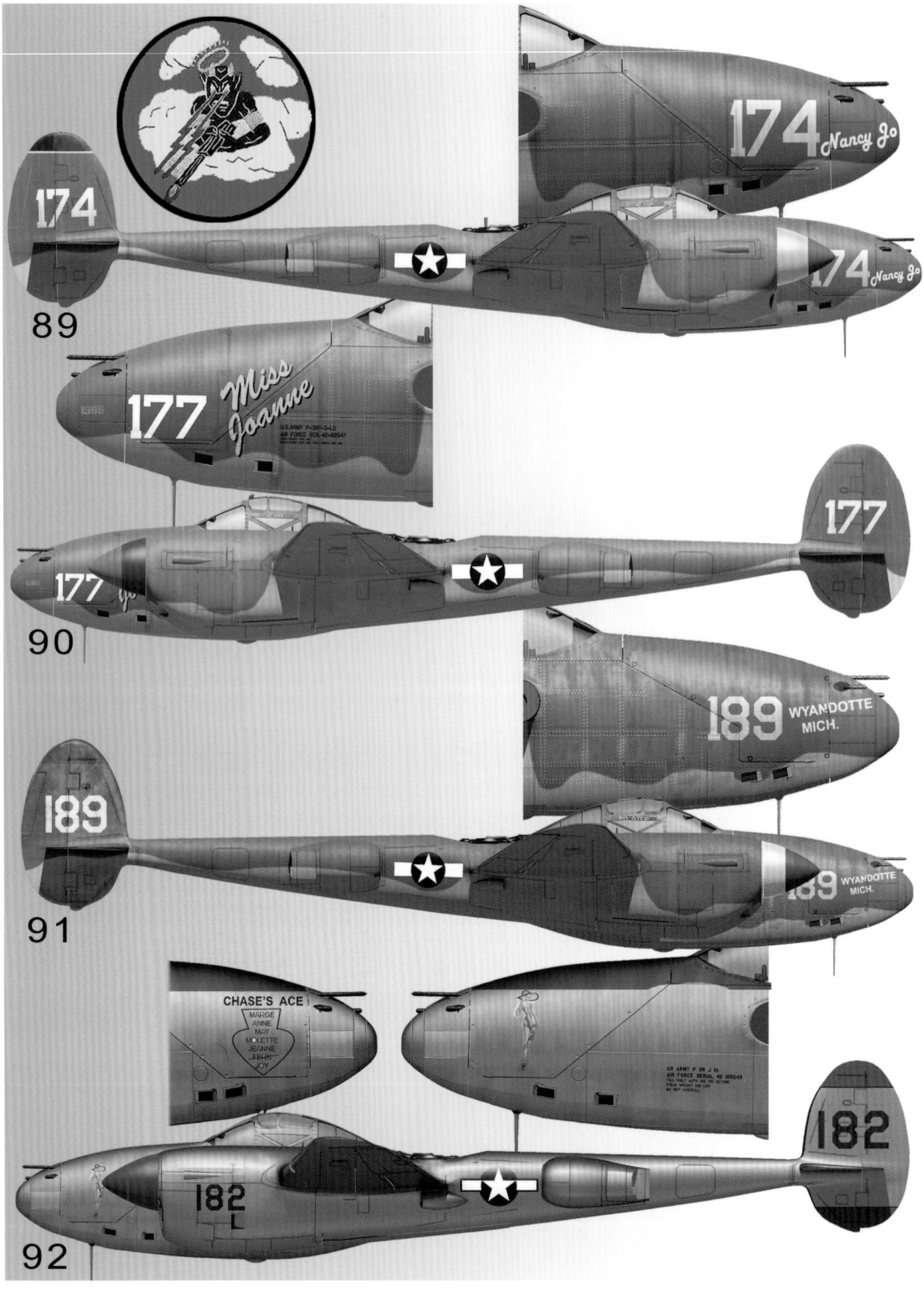
174
174 Nancy Jo
174 Nancy Jo
89
177 Miss Joanne
177
177
90
189 WYANDOTTE MICH.
189
189 WYANDOTTE MICH.
91
CHASE'S ACE
MARGE
ANNE
MAY
MOLETTE
JEANNE
FERN
JOY
182
182
L
92

433rd FS Logo

A devil in the clouds holding a machine gun with lightning bolts coming out from its eyes.

Profile 89 – P-38J unknown serial, squadron #174, *Nancy Jo*

Nancy Jo was an Olive Drab P-38J-5 photographed at Nadzab in February 1944.

Profile 90 – P-38H serial 42-66854, squadron #177, *Miss Joanne*

This fighter was assigned to and named by Lieutenant William Horton with crew chief Technical Sergeant James Guinn.

Profile 91 – P-38H serial 42-66852, squadron #189, *Wyandotte Mich.*

Assigned to Captain Jack Fisk, this Lightning was named after the small Michigan town of Wyandotte, about ten miles south of Detroit. Lieutenant Donald Revenaugh flew this fighter on 23 January 1944 from Dobodura during a B-24 escort mission. Over Wewak the P-38s fought Ki-61 Tonys and Ki-43-II Oscars. Revenaugh was last heard over the radio stating he was heading back giving his position as south of the Sepik River. He failed to return and since the weather was good it is likely he was jumped by fighters or possibly experienced a mechanical failure due to combat damage.

Profile 92 – P-38J serial 42-104494, squadron #182, *Chase's Ace*

This natural metal finish P-38J was assigned to Captain Chase Breznier at Nadzab in late March 1944, replacing his P-38H serial 42-66551 which was named *Buzzin' Cousin*. Lieutenant Joe Price belly landed this fighter at Saidor late on the afternoon of "Black Sunday".

A line up of Lightnings on the 433rd FS flight line at Nadzab in February 1944. On the far right-hand side is squadron number 174 Nancy Jo which is the subject of Profile 89.

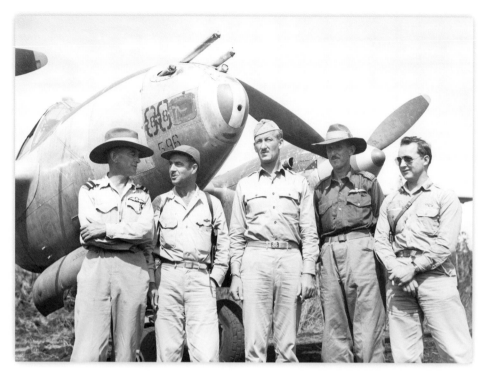

Fifth Fighter Command leadership at Tadji on 23 April 1944 (left to right): Air Commodore Frank Scherger, RAAF, Brigadier General Paul Wurtsmith, commander Fifth Fighter Command, Colonel Leif Sverdrup (chief of construction for the SWPA), Wing Commander William Dale, RAAF, and Colonel Robert Morrissey, Fifth Air Force Chief of Staff. They are standing in front of P-38J serial 42-104004 which Wurtsmith had just landed on the newly captured Tadji airstri[p.

The subject of Profile 93, P-38J serial 42-103993, at Nadzab in early March 1944 just before the photo of Bong's fiancée was glued to the airframe.

CHAPTER 22
Fifth Fighter Command Headquarters

In September 1942, Major General George Kenney structured his newly formed US Fifth Air Force so that his deputy, Brigadier General Ennis Whitehead, presided over three Commands:

- Fifth Service Command briefly under Major General Rush Lincoln then Major Victor Bertrandis. This was based at Townsville.

- Fifth Bomber Command under Brigadier General Kenneth Walker was also based at Townsville.

- Fifth Fighter Command was led by Colonel Paul "Squeeze" Wurtsmith from 11 November 1942. A few days later Wurtsmith moved his headquarters to Port Moresby.

Whilst Fifth Fighter Command Headquarters was primarily an operational and administrative umbrella for Fifth Air Force fighter operations, it developed its own detachment to which top-scoring pilots were posted. These sometimes flew on a "roving commission" basis where required. In November 1943, for example, the 348th FG commander, Colonel Neel Kearby, was transferred to the detachment. Despite being officially assigned to administrative duties, Kearby still flew combat missions.

Captain Richard Bong first served with Fifth Fighter Command in early 1944, also with a roving commission. When he returned to New Guinea in September 1944 for his third tour he was reassigned to the command as an advanced gunnery instructor, with the curious brief that he was permitted to fly missions but not to seek combat. Command fighters were serviced by the host units from which they were borrowed. Accordingly, they also carried the markings of their host units.

Bong with the photo of his fiancée on his P-38J Marge, which had been affixed shortly after the loss of Bong's comrade and friend Lieutenant Colonel Thomas Lynch on 8 March 1944 (see Profile 95). As a morale booster Bong chose to decorate his fighter in a unique fashion. Captain Jim Nichols, the 475th FG Group intelligence officer, used the nearby 6th PRG photo laboratory to enlarge a high-school photo of Margaret into a 20 x 24-inch glossy black and white print. This was then glued and varnished to the nose of Bong's Lightning. The name Marge was then decoratively painted in red.

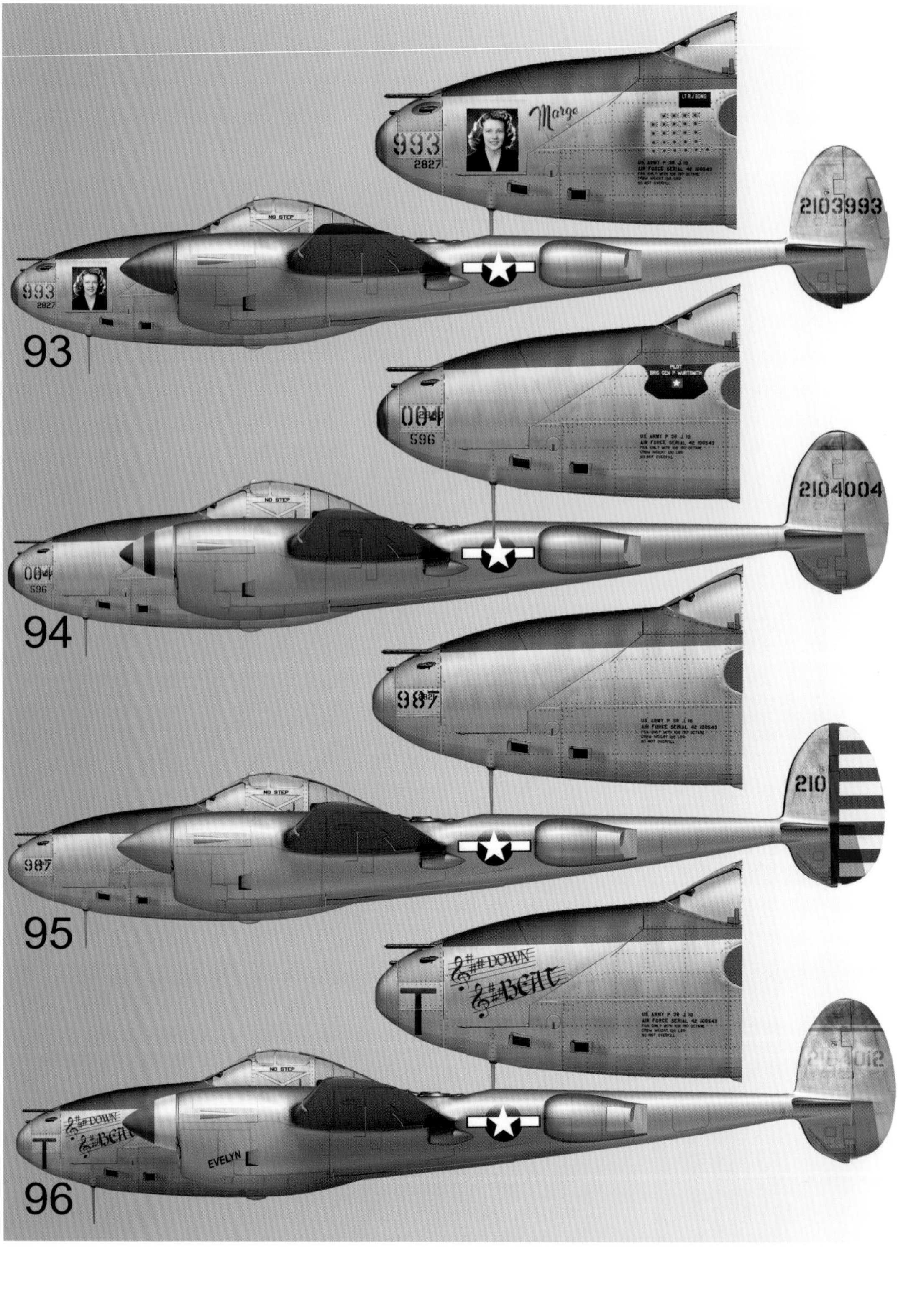
LT R J BONG
Margo
993
2827
NO STEP
2103993
93
PILOT
BRIG GEN P WURTSMITH
004
596
2104004
94
987
210
95
DOWN
BEAT
T
EVELYN
96

Profile 93 – P-38J serial 42-103993, *Marge*, Nadzab, March 1944

This Lightning is associated with the USAAF's highest scoring ace, Captain Richard Bong, and was named *Marge* after his fiancée Margaret Vattendahl. Bong was assigned this fighter when he returned to the SWPA in February 1944 for his second combat tour. It was also decorated with a photo of Margaret (further explained in the photo caption on page 117).

On morning of 24 March 1944 Bong was rostered to fly a two-ship weather reconnaissance mission to Wewak with Lieutenant Tom Malone, however Bong was pulled from the mission. Malone was then assigned Bong's aircraft for the mission, with Flying Officer Forrester as his wingman. The two departed Nadzab at 0645, but they soon ran into very bad weather and after about an hour the mission was cancelled, and they were ordered to return to base. However, after experiencing mechanical problems Malone elected to bale out from 11,000 feet. After landing in jungle, he managed to walk to a coastal plantation and was soon returned to his unit. The shattered wreckage of *Marge* was found by 32nd Infantry Division soldiers a few days later.

Profile 94 – P-38J serial 42-104004, Nadzab, March 1944

This Lightning is illustrated as it appeared shortly after assignment into Fifth Fighter Command in late March 1944. It was often flown by Lieutenant Colonel Tom Lynch and Captain Richard Bong. Brigadier General Paul Wurtsmith, the commander of Fifth Air Force Fighter Command, flew this fighter into Tadji on 23 April 1944 - the first Allied fighter to land there following its capture.

Profile 95 – P-38J serial 42-103987, Nadzab, March 1944

On 9 March 1944, this brand-new Lightning crashed near Aitape airfield into a mangrove swamp where its wreckage lies today. On this day Captain Bong and Lieutenant Colonel Thomas Lynch flew together, strafing luggers anchored off the coast. IJN No. 90 Garrison Unit was ensconced nearby, equipped with two light 7.7mm machine guns and three heavy 13mm "machine-canon". During a run Bong witnessed Lynch's P-38 take hits from the gunfire, with the entire nose section blown off and the starboard engine set afire. Lynch climbed to 2,500 feet to bale out, but apparently experienced difficulty. At just 100 feet the Lightning exploded, ejecting Lynch whose parachute streamed, and he fell to earth not far from the wreckage of his aircraft. The same gunners hit Bong too; he had to feather an engine to return to Nadzab where his crew chief counted 78 bullet holes. The markings are taken from the wreckage at the site photographed in 2020, thus enabling Lynch's fighter to be accurately illustrated for the first time.

Profile 96 – P-38J serial 42-104012, squadron letter T, *Down Beat*, Nadzab, April 1944

This Lightning was transferred to Fifth Fighter Command in April 1944 from the 80th FS which had named it. It later proceeded to the Philippines theatre.

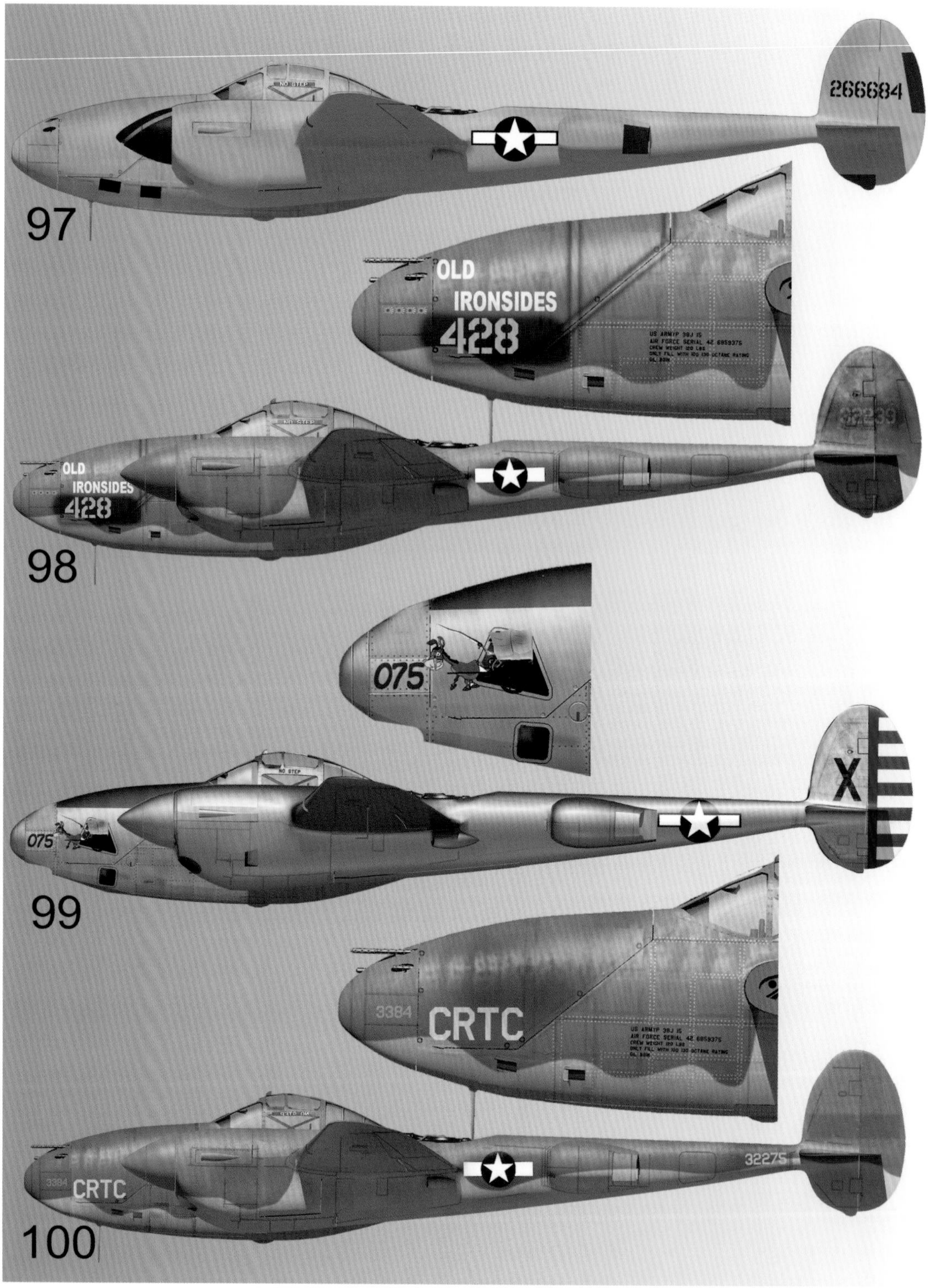
266684
97
OLD
IRONSIDES
428
98
075
X
99
3384
CRTC
32275
100

CHAPTER 23
Unusual Lightnings

These Lightnings have been selected because of the unique role they played in the theatre.

Profile 97 – P-38H 42-66684, 18th FG

Following service with the 339th FS, this P-38H had all armament removed and was painted overall yellow to make it highly visible. It was then used to tow gunnery targets at Guadalcanal for the 44th and 70th FS when they took delivery of their natural metal finish Lightnings in March/April 1944. Note the serial number was reapplied in black on the fins.

Profile 98 – P-38G serial 43-2239, squadron #428, *Old Ironsides*, 44th FS,

Originally assigned to the 339th FS as squadron number 138, on 29 March 1943 70th FS pilot Lieutenant Sam Howie badly damaged the aircraft when he force-landed at Guadalcanal following an attack against the floatplane base at Faisi. The airframe was towed to the boneyard however 44th FS engineers found sufficient spare parts to painstakingly restore the airframe to airworthy condition. The fighter, used as a hack, is illustrated as it appeared at Fighter Two in March 1944. Shortly afterwards there was an altercation with 339th FS engineers who tried to reclaim the aircraft for their own squadron's use.

Profile 99 – F-5A serial 42-13075, 25th PRS

This reconnaissance Lightning was retired from combat by the 25th PRS and used as a hack by the parent 6th PRG. The letter X on the tail signified it was no longer in active service, and the artwork of a donkey pulling a cart reflected its utilitarian status, painted over the top camera portal. The aircraft was taken off charge on 6 October 1944.

Profile 100 – P-38G serial 43-2275, CRTC, Port Moresby March 1944

The first G model P-38s were assigned into the Fifth Air Force inventory in early 1943. This airframe showcases markings for one of the last P-38s to operate in New Guinea. First operated by the 80th FS, this airframe was transferred to the Combat Replacement Training Center, which operated under the 360th Service Group, in early 1944 at Nadzab. During a navigation exercise to Port Moresby on 25 March 1944, it crashed on the return take-off at Pioneer's Crossing, located at the southeast end of Seven-Mile 'drome. Note that CRTC repainted the serial number to the rear of the boom after the unknown 80th FS alphabetical identifier and original serial were over-painted on the fin. The Constructor's Number 3384 remained extant on the nose in yellow stencils.

The subject of Profile 97, P-38H 42-66684, at Guadalcanal shortly before the serial number was reapplied.

The subject of Profile 98, P-38G serial 43-2239, on Guadalcanal in March 1943 when serving with the 339th FS.

Sources

Research for this volume draws only on primary sources. The author's extensive collection of photos and notes from many years of field trips contains a plethora of information. It is impractical to list these minutiae, other than to cite the main sources below.

Website www.pacificwrecks.com and its hard-working owner, Justin Taylan
Headquarters Fifth Air Force Special Orders No. 272, 29 September 1943
Allied Translator and Interpreter Section (ATIS) Reports
Allied Air Force Intelligence Summaries (Australian War Memorial)
ANGAU patrol officer reports of Allied crash sites 1940s–1970s
Lockheed Aircraft Corporation Engineering and other historic Records
P-38 markings details from relevant Individual Deceased Personnel Files (IDPF)
Field Trips by author throughout New Guinea and the Pacific, 1964–2017, including those with James Luk around Lae in 1976
Pacific Aircraft Historical Society – Wreck Data Sheets
PNG Colonial Office – Civil Administration Records and PNG Cultural Museum
Aircraft Movement Entries, Townsville Control Tower, 1942–43
Field notes of Robert Greinert and John Douglas, wreck sites, PNG, 2011–2019
Papua New Guinea Catholic Mission Association, Field Trips of, Papua New Guinea

Microfilms/ official records

Fifth and Thirteenth AF Units via Maxwell AFB: 5th Air Force Establishment, Fifth Fighter Command, 8th FG, 8th PRS, 25th PRS, 26th PRS, 35th FG, 36th FS, 80th FS, 4th Air Depot, 27th Air Depot, 12th FS, 36th FS, 39th FS, 67th FS, 68th FS, 70th FS, 339th FS, 418th NFS, 419th NFS, 421st NFS, 431st FS, 432nd FS, 433rd FS, 475th FG, 6th PRG and 18th FG.

Pluto, a P-38H assigned to the 12th FS, patrols over Santa Isabel in the Solomons in late 1943. The communicative artwork of Pluto the reluctant "dawg" gave visitors to the squadron's flight line light relief.

Index of Names